INTERPET
HANDBOOKS

YOUR
HEALTHY
GARDEN POND

YOUR
HEALTHY
GARDEN POND

STEVE HALLS

INTERPET PUBLISHING

© 2000 Interpet Publishing,
Vincent Lane, Dorking, Surrey,
RH4 3YX, England.
All rights reserved.
ISBN: 1-903098-08-4

Credits
Created and designed: Ideas into Print,
New Ash Green, Kent DA3 8JD, UK.
Artwork illustrations: John Sutton
Computer graphics: Stuart Watkinson
Production management: Consortium,
Poslingford, Suffolk CO10 8RA, UK.
Print production: Sino Publishing
House Ltd., Hong Kong.
Printed and bound in China.

The author
Steve Halls has worked in the aquatic
trade all his working life. His present
post of brand manager for a range of
leading water gardening products brings
him into close contact with the
practicalities of maintaining a healthy
environment for aquatic plants and fish.
This 'hands-on' involvement with the
needs of pondkeepers provides him with
the ideal platform to explain clearly and
accurately how to create and sustain a
successful water garden.

Below: Koi in crystal clear waters.

Contents

Note: Throughout this book, capacities are quoted in litres. To convert litres to imperial gallons, multiply the number by 0.22. To convert litres to US gallons, multiply the number by 0.26.

The stress-free way to a healthy pond

The sound of trickling water and the flash of colour provided by bright, healthy fish is the aim of most pondkeepers. A healthy pond with clean water and lush plant growth will brighten any garden, large or small, and bring hours of pleasure and relaxation. It is not difficult to achieve such a pond, given the correct information, some planning and a little work. This book aims to make achieving a healthy pond a relatively stress-free exercise.

Understanding the environment in which fish and plants thrive is a key element in maintaining a healthy pond. We begin by looking at water quality, how to achieve it, why it is important and how to rectify problems. The life support system of most ponds is artificial filtration and there is information to help you choose a filter and pump suited to your pond. Green water caused by algae is the pond-owner's most common cause for complaint, and we cover all the latest methods of algae control and prevention. Pond plants are not only aesthetically pleasing, they also play an essential part in the healthy pond. The chapter on pond plants will help you keep pond plants in excellent condition all year.

The next section offers guidance on selecting, transporting and introducing fish to the pond, followed by a brief survey of some suitable pond fish. In the chapter on fish health, you will find advice on identifying the most common ailments, their prevention and treatment. From time to time it will be necessary to give the pond a complete overhaul and this is explained in detail. The book concludes with a guide to seasonal maintenance.

WATER QUALITY

Keeping a healthy pond, full of aquatic life, is a matter of a little work and an understanding of the basic principles involved in maintaining good water quality. If we look after the water in our ponds, the water will look after the fish.

The quality of the water is the key to keeping a healthy pond. Your fish and plants depend on you for their well-being. Your pond is a closed environment; the fish live in the water into which they excrete and in which plant matter decays. It is therefore vital to understand the biological activity that takes place in your pond and how, through filtration, you can greatly assist this essential activity. In this section, we look at three major aspects of water quality: the nitrogen cycle, the acid-alkaline balance and oxygen levels.

Moving water

A balanced, healthy pond will include fish, plants and moving water to provide a feature suitable for any garden throughout the year.

Filling and topping up the pond

Tap water is usually used for filling and replacing water lost through evaporation. It has chlorine added to it to reduce bacteria levels and ensure that it is safe for us to consume. However, chlorine is highly toxic to aquatic life and must be removed before it can harm life in the pond.

Understanding your pond water

Water in the pond is never pure H_2O. Rather, it is a combination of minerals and compounds that, when maintained correctly, provide an ideal environment for fish and plants.

The main properties that concern the pondkeeper are firstly the nitrogen cycle (ammonia, nitrite and nitrate), and secondly the acidity or alkalinity of the pond water. If you understand the influence of these factors on the water quality, it will make maintaining a healthy pond much easier.

The nitrogen cycle

It is important to appreciate that a pond differs from natural bodies of water in two fundamental ways. Firstly, ponds do not receive a constant supply of fresh water to replace the 'lived in' water that they contain. Secondly, most artificial ponds have a far higher stocking level and bioload (the sum of all the biological activity going on) than natural ponds and lakes. For these reasons, you as the pondkeeper must be aware of the nitrogen cycle and how to maintain it at a healthy equilibrium.

Nitrification

Nitrification occurs when highly toxic ammonia is converted by beneficial bacteria in a filter into less harmful nitrite. The fish in your pond will naturally produce ammonia as they excrete. Uneaten food and decomposing plant life add to the level of ammonia, which if left, can build up to lethal levels for

Above: Using a tap water conditioner is simply a matter of diluting the required amount in a watering can full of water and pouring it over the pond surface.

Fortunately, this is easy to do, either with a tap water purifier or, more commonly, with a proprietary dechlorinator available from aquatic shops and garden centres. Be sure to add the correct amount of dechlorinator, both when you first fill the pond and whenever you top it up. This is especially important in the summer months, when the rate of evaporation is higher and chlorine levels are often increased in tap water to cope with increased bacterial activity. Failure to remove chlorine will cause great distress to fish, and could even kill them, as the chlorine affects their gills and strips the layer of mucus on the body that protects fish from disease.

The nitrogen cycle

*This illustration shows
how nitrogen circulates in
the pond. The bacteria that
convert one nitrogen-
containing compound to
another occur naturally.
To prevent ammonia
and nitrite building
up in the pond, it
is essential to
encourage
these bacteria
to thrive in filters.*

*When fish eat plants
the digestive processes
break down the
proteins and create
ammonia as one of the
waste products. Some
of this is excreted from
the gills.*

*Plants absorb
nitrates as 'food' and
the nitrogen is used
in the construction
of plant proteins.*

*Aerobic bacteria
oxidize the nitrite
to nitrate (NO_3).*

Ammonia (NH₃) is
present in fish excretions
and also builds up in the
water as plants and
uneaten food decay in
the pond. It is highly
toxic to fish.

Aerobic bacteria
oxidize the ammonia
to nitrite (NO₂).

Small versions of this cycle
appear next to each test
shown on the following
pages with the relevant
part highlighted.

pond livestock. When the process of nitrification has oxidized ammonia to less harmful, though still toxic nitrite, a different type of bacteria take over and convert nitrite into relatively harmless nitrate. This process is a further stage in the nitrogen cycle, and also takes place in the pond filter.

Ammonia levels
Ammonia (NH_3) tends to occur mainly in new ponds, where the filtration system has not had time to build up the beneficial bacteria required to break down excreta, food and decaying plant matter. Ammonia levels above 0.025mg/litre can be lethal to delicate species of fish. It is therefore important to monitor ammonia levels in the first few weeks of introducing fish to the pond.

Nitrite
Checking nitrite (NO_2) levels is often seen as the most important regular test in a pond. Nitrite is highly toxic to livestock and can occur through overstocking, overfeeding or lack of filtration. The ideal nitrite level is zero, and anything above 0.1mg/litre will harm fish.

Testing your pond water
It is recommended that you test your pond water on a weekly basis. This will alert you to any problems, such as overstocking, failing filtration or overfeeding, and can prevent expensive and upsetting losses of fish and plants.

Water tests are inexpensive and simple to carry out. Test kits are usually found either in tablet form or as liquid reagents. Both types involve

Testing for ammonia

Ammonia is most commonly detected in new ponds, where the bacteria have not reached sufficient levels to break down the organic waste in the water.

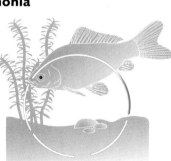

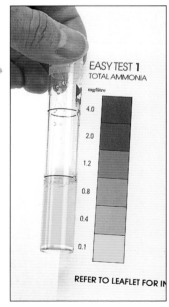

Right: This test involves adding two tablets to a sample of pond water and comparing the colour to a printed chart. Ammonia readings above zero indicate a problem, which, if left unresolved, can cause fatalities. Repeat the test every day until it is clear.

WATER QUALITY

Testing for nitrite

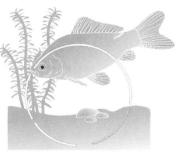

In the second stage of the nitrogen cycle, bacteria in the filter (and occurring naturally in the pond) break down ammonia into nitrite, which is less toxic.

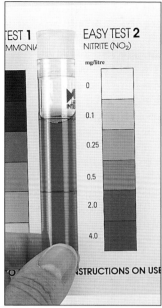

Right: *In this test, you only need to add one tablet to the water sample and shake the tube. The colour takes ten minutes to develop and then you can compare it to the printed chart. Nitrite levels of above 0.1mg/litre are toxic to fish and need remedial action. (Here the reading is very high, approaching 2mg/litre). Causes of high nitrite levels may include lack of filtration, overstocking or overfeeding.*

Testing for nitrate

In the third stage of the nitrogen cycle, nitrite is converted by bacteria in the filter into nitrate. Plants absorb the nitrate and use it as a food source.

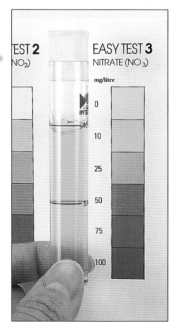

Right: *This is another two-tablet test and the reading here shows about 10mg/litre of nitrate in the sample of pond water. This level is not a dangerously high one. Although relatively harmless, high nitrate levels may lead to unsightly algae growth and green water. If the pond has a good growth of aquatic plants, particularly submerged ones, then they will use up much of the nitrate produced in the nitrogen cycle and prevent algal blooms and distress to the fish.*

comparing a colour change in the treated sample to a printed chart.

Acidity and alkalinity

The acidity or alkalinity of water is a measure of the quantity of hydrogen ions available. (The H_2O water molecules split up into positively charged hydrogen ions H^+ and negatively charged hydroxyl ions OH^-.) The levels of acidity and alkalinity are shown as pH, literally the 'power of Hydrogen'.

The pH scale ranges from 0 to 14, with the lower end of the scale being the most acidic. The middle point of 7 is neutral and readings above this mean the water is alkaline. Extremes at either end of the scale are lethal to livestock, and you should aim to keep the pH within 7-8.5 in order to maintain a healthy balanced pond. The pH scale is logarithmic, so a pH of 8 is ten times more alkaline than a

pH of 7. Do not adjust the pH by more than 0.5 of a unit at one time.

Symptoms of high pH levels

If you do not adjust high pH levels, plant life may begin to suffer. Growth rates will slow down and oxygenating plants, such as *Elodea crispa*, may take on a white powdery coating. This is caused by calcium precipitating onto the plant's leaves. High pH levels also make ammonia more toxic and fish begin to show signs of distress. They may die with no external symptoms or signs of infection. Filters become less effective in high pH conditions and nitrite levels may rise in the pond.

The causes of high pH levels

New ponds are often too alkaline. This sometimes happens because lime from concrete leaks into the pond. Be sure to wash a new pond thoroughly

Testing for pH

Left: Simple, accurate kits are available for testing pH levels. This one involves adding one tablet to a sample of water and shaking vigorously. Do not put your finger over the end because this may affect the test result.

Right: Always follow test kit instructions and wait for the time specified before comparing results with the colour chart. Ideally, pond water should be pH 7-8.5.

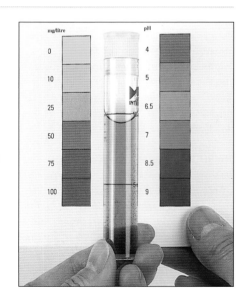

before filling and stocking it with fish to ensure that dust from surrounding paving or concrete blockwork are not present in the pond. If the pond is constructed from concrete, apply a good-quality pond paint to water-proof the concrete and prevent toxic lime leaching into the pond.

Excessive levels of algae growth can also cause high pH. As the algae absorbs CO_2 the pH rises. At night, when the level of CO_2 builds up again, the pH of the pond water is restored to a safer level.

Dealing with high pH (alkalinity)

Make sure that all concrete blocks or paving slabs are sealed with a waterproofer. Use solvent-based sealers or acrylic pond paint. Remove any small pieces of concrete from the pond that may have fallen in during construction. Remove blanketweed and reduce the level of algae in the pond (see pages 38-49)

If using an algicide treatment, it is a good idea to carry out a partial water change to dilute the effects of algae dying off in the pond. Before treating for blanketweed problems, remove as many of the green strands as possible.

Use a proprietary pH adjuster to establish an equilibrium in the pond. Follow the instructions carefully and remember that sudden changes in pH should be avoided.

Symptoms of low pH levels

If the pH level is too low, fish may become listless and their fins may take on a bloodshot appearance. Pond snails will begin to die off,

because their shells rely on the calcium provided by more alkaline water. Many plants will show slow rates of growth. *Egeria densa* and other oxygenating plants, for example, are particularly prone to pH extremes. In acid conditions, biological filters will work less effectively, possibly causing nitrite levels to rise in the pond.

The causes of low pH levels

Water entering the pond through soil or after heavy rain may be softened and acidified by humic acids. These natural acids occur particularly if the surrounding area is peat-based or if the pond has decaying plant matter, either from wind-blown leaves or from pond plants dying back. Rainfall from polluted areas is also often acidic and heavy showers can significantly reduce the pH in a small pond, as well as introducing toxic substances. Fish excreta will also tend to acidify the water as it breaks down. For this reason, it is important to test the pH on a regular basis, as pond water will have a tendency to acidify over time.

Dealing with low pH (acidity)

Changing up to 20% of the pond water volume will help to prevent pH drops, as the introduction of new water will dilute the acidity caused by decomposing plants and fish excreta. Regular pond maintenance, including the removal of leaves and vacuuming the bottom of the pond to prevent a build-up of detritus, will also help to maintain a stable pH. In autumn and winter, it is a good idea to install a pond cover net to prevent leaves from falling into the pond. It is also important to remove dead and dying leaves in and around the pond. If problems occur in the pond as a result of low pH, consider using an alkalinity buffer to maintain the correct balance.

Low oxygen levels

During periods of high temperature, many pondkeepers experience fish losses for no apparent reason. More often than not, this is caused by lack of dissolved oxygen in the water.

High temperatures combined with a lack of water movement can cause the level of gaseous exchange in the pond to drop to dangerously low levels. Gaseous exchange is an essential part of a healthy pond, because it enables toxic gases, such as carbon dioxide, to leave the pond in exchange for oxygen. Fortunately, the problems caused by low oxygen levels in the average garden pond are easy to recognize and rectify.

Symptoms of low oxygen levels

In low oxygen conditions fish become listless and can often be seen gasping at the surface of the pond. Remember that beneficial plants can introduce oxygen throughout the day, but will cease to do so at night. This may be the reason for any unexplained fish losses overnight.

Fish such as golden orfe and rudd are often the first to suffer, as they require higher oxygen levels in the water than most other species. In extreme cases of depleted oxygen, the water may give off a nasty odour.

The causes of low oxygen levels

Low oxygen levels in the pond may be explained by a number of reasons. Fountains and waterfalls normally in use may have been turned off, for example, and warm, humid weather naturally reduces the dissolved oxygen content of the water. The

Left: During hot periods it is advisable to increase oxygenation. Using an airpump connected to airstones creates a rising stream of bubbles that agitates the water surface and helps to raise oxygen levels.

pond may be overstocked or the fish overfed, causing excess food and waste to decompose and use up oxygen. Algal blooms, blanketweed and green water may be using up oxygen at night. The pond surface may be covered with plant life; if there are too many lily leaves or floating plants, such as duckweed, gaseous exchange cannot take place.

Dealing with low oxygen levels

Oxygenation of pond water will occur wherever the surface of the pond is disturbed and comes into contact with air. At higher temperatures, less oxygenation takes place, so it is important not to overstock the pond in spring (i.e. before the pond water has fully warmed up), when oxygen levels are higher. In summer, make sure that fountains and waterfalls are left running all night. Alternatively, buy a pond airpump and run this all night. The stream of bubbles from the pump's airstone will break the water surface and helps to increase the oxygen content of the pond

considerably. When buying an airpump, choose one that will pump air from at least halfway down the pond's depth to ensure reasonable water circulation, otherwise there may be a lack of oxygen at lower levels in the pond. Do not turn off pumps and filters at night. Leave filter systems running 24 hours a day.

Remove excess duckweed and other surface-growing plants from the pond. The more surface area available, the easier it will be for oxygenation to take place. For the same reason, do not allow algae levels to build up. If green water occurs, use a proprietary algae treatment to clear it or consider installing an ultraviolet sterilizer (see page 42).

If the pond is overstocked, perhaps because fish have bred, consider giving away some of fish or uprating the filtration system.

Finally, do not allow excess levels of silt and decaying plant matter to build up in the pond. Use a pond vacuum to clean the base of the pond on a regular basis.

Right: Fish such as these rudd and golden orfe need good water movement and high levels of dissolved oxygen to thrive. They will be among the first fish to suffer during hot weather in ponds with a lack of water movement and therefore low levels of dissolved oxygen.

POND FILTRATION

Pond filters are an essential part of all but the most balanced wildlife ponds, where fish stocking levels are extremely low. Here we explore the types of filter and the benefits of filtration, rather than discuss specific models.

Crystal clear water

A glass panel forming one side of a pond provides a stunning view of crystal clear water teeming with fish – not possible without filtration.

There are two principal forms of pond filtration. The first is to physically remove any solids and waste from the pond. This is known as mechanical filtration and is the first stage in any multistage filter. There are several types of filter media that can be used to trap solids and these will be discussed later.

The second form of filtration is biological and works by harnessing naturally occurring aerobic bacteria that break down pond waste. Bacteria will break down the ammonia present in fish excreta and

generated by decaying plant matter and convert it into nitrite. A different type of bacteria then take over and feed on the nitrite, converting it to less harmful nitrate, which can be used by plants as food.

In addition to mechanical and biological filtration, some filters incorporate chemical filter media, such as zeolite and activated carbon. As the name suggests, chemical filtration removes impurities by using chemicals, rather than bacteria.

Pond filters require a pump either to push water into them or pull water through them. As water enters the filter, it is processed by a series of media that become increasingly smaller. These eliminate the finer solids as water passes through them. Do not use a fine filter medium straightaway as it will quickly clog and require constant maintenance.

Types of pond filter

There are many types of pond filter and they all perform the same basic task in one way or another. The object of a filter is to provide clean, healthy water for your fish and plants to thrive in.

Filters are generally rated in terms of the maximum volume of pond water they will filter. Take care when buying a filter, as in many cases the rating is based on very low stock levels, on the prevailing conditions in certain climates (warmer conditions need bigger, more efficient filters), and on a very strict feeding regime. As most pondkeepers will exceed the low stock levels suggested with the filter, and will therefore add more food than is allowed for, you should consider buying a filter at least one size larger than the minimum suggested size. If in any doubt, seek the advice of a reputable dealer.

The most basic type of filter will be a block of sponge that fits onto your pond pump. This is suitable for small ponds that hold only a few

Above: A multichamber filter system purpose-built for a koi pond. Koi enthusiasts are keen to create their own filter systems in the pursuit of perfect water conditions for these prized fish.

hundred litres and with low stocking levels, but not for larger ponds or those with high numbers of fish.

The sponge performs two tasks. Firstly, it will trap solid material and prevent the pump from clogging, which would reduce the effectiveness of a fountain or waterfall. Secondly, the sponge will provide a breeding ground for the beneficial bacteria that break down ammonia and nitrite (see pages 14-15).

Sponge filters are limited in their capability for a variety of reasons. They do not have a particularly large surface area and, of course, all the waste stays in the pond. A better way to filter the pond is to use an external filter of some kind.

External pond filters

The simplest external pond filters are box filters that contain a variety of media. Water is pumped from the pond into the filter and passes through the different media before re-entering the pond via a return pipe or waterfall. These external filters are simple to maintain and well suited to small and medium-sized ponds. When designing your pond, it helps to incorporate the siting of the filter in the initial planning. Most box

planning. Most box filters are not pressurised and must therefore be placed at the top of a waterfall so that water can flow freely out of the filter, relying on gravity to take it back into the pond, rather than being pumped up into the pond.

Pressurized filters are more versatile than basic box filters, and there are a number of types to choose from. All have a sealing ring, which makes them watertight. Water can be pumped into them and then forced out under pressure. This enables the filter to be placed below a waterfall or used with a fountain jet to provide a dual-purpose filter.

A basic external box filter

The water flows down through the bristles of these brushes as a first stage mechanical filtration.

After the initial straining process, the water passes under a plastic partition and up into this larger compartment filled with plastic tubes. Here, beneficial bacteria convert ammonia to nitrite and nitrite to nitrate in the biological filtration stage.

Water is pumped from the pond and enters the filter here.

Water returns to the pond.

A multistage external box filter

Water enters from the pond and swirls around the cyclone chamber, where heavy particles settle out.

Water flows down through successively finer grades of foam that strain out smaller particles. For new or dirty ponds, fit a filter brush.

Once it has reached the bottom of the container, the water rises up through a circular bed of highly porous aquarock, where bacterial colonies carry out biological filtration.

Water returns to the pond by gravity.

Multichambered filters

Multichambered filters are available for more comprehensive filtration systems. They consist of three or four separate compartments connected together. Each compartment usually houses a different filter medium and normally has a drain tap for ease of maintenance.

The different stages of filtration

The first stage of a filter may consist of brushes or large plastic tubes, both of which will trap large particulate waste without clogging and allow water to continue its passage through the filter for further processing.

Alternatively, you can use a vortex chamber. Vortex chambers are usually conical in shape, with an inlet at the top and the outlet at the bottom. Water enters the vortex parallel to the internal wall of the vortex chamber. The resulting spiralling effect – caused by water circulating around the inner wall and slowing as it gets nearer the centre of the chamber – causes dirt and debris to settle at the bottom of the chamber. From here it can be discarded via a drain valve.

The second stage of filtration may be sheets or blocks of foam, which trap smaller particles of waste and allow cleaner water to enter the biological part of the filter. The final stage of a simple filter will be the biological one, incorporating a

A simple pressurized filter

Optional fountain head

Screw lid with 'O' ring to retain water pressure.

Water is pumped into the filter here.

This pressurized filter can be used in-pond to provide a fountain or externally to supply a waterfall. The filter works both mechanically and biologically, using foam, aquarock and filtroc.

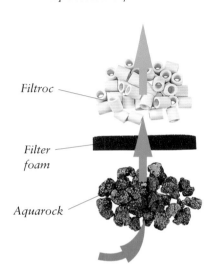

Filtroc

Filter foam

Aquarock

medium with a very high surface area. The more surface area is made available, the more bacteria will be present to break down toxic ammonia into nitrite and then into nitrate, before the water is returned to the pond.

In addition to these beneficial bacteria, a healthy, well-run filter may accumulate water fleas (*Daphnia*) and bloodworms that will aid the filtration process by feeding from the waste in the filter chamber.

Getting the most from the filter
If the filter is to operate efficiently you must follow a few basic rules. Be sure to leave the filter running 24 hours a day. Failure to do this

will cause the beneficial bacteria to die off, as they need a constant supply of oxygenated water to keep them alive. If the filter stops running for any reason, drain the water out before restarting it, as any water retained in the filter will contain a high level of dead bacteria.

Filter media
There is a wide variety of filter media available today and the choice can be somewhat confusing. Here we look at some of the more popular options. Bear in mind that no single medium is ideal for all applications; some may be good at trapping solid waste, while others may have a high surface area for bacteria to grow on.

Ideally, you should use a variety of media in a filter to maximize its efficiency and effectiveness.

Gravel is probably the cheapest filter medium available. It has a reasonable surface area and can be used for biological filtration. However, it is heavy, not easy to clean and prone to clogging, which causes 'channelling'. This is the term used to describe a filter that has blocked areas. In these cases, the water is 'channelled' through the filter where the flow is less restricted. This means that only a portion of the filter is being used, thus reducing its efficiency. If you use gravel, stir it regularly to keep it from compacting.

Filter foam is widely used and available in different grades of coarseness. The foam used for pond filters is open-celled and non-toxic. Do not be tempted to use cheaper foam, such as that found in furniture, as this will leach toxins into the water and poison the fish.

Filter foam has a large surface area and is easy to handle, although it does have a tendency to dump dirty water back into the filter when it is disturbed for cleaning. Filter foam is relatively expensive and will need replacing every two or three years. It can be used for both mechanical and biological filtration. One drawback is that very fine foam has a tendency to clog quickly.

Clay granules, as used for garden pond filters, have been heated to high temperatures, producing a relatively

lightweight, round filter material that is used both for mechanical and biological filtration. Because of their shape, they are less prone to clogging than gravel and allow a good water flow. Their reasonably high surface area makes them suitable for biological filtration and they are frequently incorporated into filters using only one medium. Like gravel, they are not particularly easy to remove and clean on a regular basis.

Plastic tubes and balls are just some of the plastic filter media available. Most are the 'hair roller' type, a 5-7.5cm (2-3in) tube of plastic, often corrugated. This medium is extremely lightweight and eminently suitable as a prefilter. The high flow rate through this medium and its ability to trap solids without clogging make it an ideal first or second medium in a multichamber filter.

Filter brushes are extremely popular as mechanical filters, as they allow a high flow rate to pass through them without clogging, while efficiently trapping large solids. They are easy to clean and can be removed from the filter and hosed off.

Many multichamber filters use brushes for the initial mechanical filtration. There are many grades of filter brush available and quality is often reflected by price. Good-quality brushes will be dense and robust; poorer-quality brushes may only last for one or two seasons.

Aquarock (Alfagrog) is an extremely porous filter medium and ideal for biological filtration due to its large surface area. To get the best results from this medium, place it in one of the last chambers of a filter, where the water has already been prefiltered. This prevents clogging.

Remember that when you clean any biological filter medium, it is important to rinse it in water taken from the pond, rather than in tap water. This prevents harmful chlorine affecting the bacteria living on the filter media.

Filtroc is a sintered glass medium with many thousands of microscopic cavities. (The sintering process involves heating the material to a high temperature to create the texture.) Because filtroc has a vast surface area, it can be used as one of the final stages in a filter and provides an ideal breeding ground for beneficial bacteria. Although it may appear expensive, a filter requires far less filtroc than other types of biological media due to its large surface area. To make it easier to clean you can place filtroc into a mesh bag.

Matting is a versatile filter medium made from fibres coated in latex. Its high surface area and coarse design makes it an excellent, if expensive, all-round filter medium. It is easy to wash and lasts for several seasons.

Zeolite is an aluminium-silicate material, available in grades from fine through to large pieces. It is best used on a temporary basis to aid bacteria in controlling the ammonia

Filter media

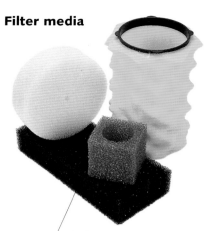

Foam is lightweight and easy to use, mainly as a mechanical filter.

Filter brushes are durable, easy to clean and excellent for trapping large waste particles.

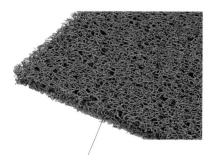

Filter matting is easy to maintain and suitable for both mechanical and biological filtration.

Clay granules are lighter than
gravel, with a larger surface area
for bacterial growth.

Activated carbon will adsorb
waste and toxins. Discard
once it is 'full'.

Gravel is cheap and provides a
reasonable surface area for
mechanical and biological filtration.

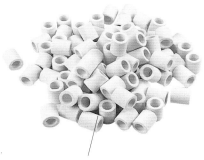

Filtroc has a very large
surface area and is ideal for
biological filtration.

Aquarock is an ideal biological filter
medium, being highly porous and
with a very large surface area.

Plastic tubes and balls allow
good water flow while
trapping waste.

level in a new pond, after a power cut, or after the introduction of new fish. Zeolite's advantage over activated carbon is that it is rechargeable. To recharge it, simply mix cooking salt with water at a rate of about 6gm per litre. Soak the zeolite for 24 hours, then rinse it thoroughly in fresh water, before placing it back in the filter.

Never add zeolite if salt is being used as a treatment, as it will cause the zeolite to discharge ammonia back into the pond.

Activated carbon can be used to aid the nitrogen cycle and to remove discoloration from the pond. Carbon can also be used after medication to remove any remnants of the chemicals used in the treatment. Activated carbon cannot be recharged; change it frequently so that the adsorbed chemicals are not released back into the pond.

Choosing a filter
When choosing a filter, select one that will easily cope with the size of your pond. The rating of some filters is based on very few fish and a low feeding regime. With a few exceptions, size is important. If two filters are both rated for 4500 litres, the larger one will generally filter the pond more efficiently, assuming that

An in-pond filter system

Inflowing water first passes through a foam sleeve for mechanical filtration.

Tethered float enables you to locate the filter for maintenance.

The inner compartment houses a mesh bag of porous biological media.

Three grades of foam sleeve: fine (yellow), medium (blue) and coarse (black).

Filtered water drawn away by pump.

Cleaning a foam filter insert

1 Although the foam in a multistage pond filter is not carrying out the major biological filtration, it is advisable to collect some pond water to wash it in to keep alive any bacteria it harbours.

2 This is the foam sleeve from the in-pond filter shown opposite. To clean it simply plunge it into a bucket of pond water and squeeze it a few times to rinse out the accumulated sludge.

their design and the filter media are similar. With the wide range of filters on the market today, you are strongly advised to seek the advice of your local aquatic and water gardening specialist to help you make the correct choice.

Maintaining the filter
Even the best filters require regular maintenance to function properly and there are a few simple rules to follow. Clean mechanical filter media regularly to ensure that the flow rate through the filter does not become restricted. Rinse brushes under a

running hose or tap and remove bags of mechanical media for cleaning in the same way.

If possible, buy a filter with a slide valve or drain tap, which makes it simple to remove heavy waste. Open the valve for long enough to allow waste from the bottom of the filter to drain away. This is particularly important for filters where a vortex is used to trap heavy solids.

Never wash biological filter media in fresh water, as this will result in the loss of the beneficial bacteria that help maintain a healthy pond. When cleaning biological media – or any media in a single chamber filter – simply stir the media through to disturb the debris and flush the dirty water away to waste. Alternatively, remove the media from the filter and rinse it in a bucket of pond water. Filters should be routinely cleaned throughout the season, particularly if the flow rate drops noticeably. Do not completely clean out your pond filter during spring and summer unless it is absolutely necessary. If a complete overhaul is required, first remove the biological media to ensure that there is less chance of the bacteria dying. If necessary, thoroughly clean out filters and replace the media in the winter months, when biological activity in the pond is at its lowest.

Calculating the volume of your pond

Measure the average length, width and depth of your pond in metres and multiply them together. This gives the capacity in cubic metres. To convert this to litres multiply the answer by 1000.

Example:
Average length: 2.2m
Average width: 1.5m
Average depth: 0.9m

2.2 x 1.5 x 0.9 = 2.97 cubic metres
2.97 x 1000 = 2970 litres

A typical pond pump

This robust amphibious style pump can be used either in or out of water.

Fountain head

The prefilter housing keeps the sponge filter in place and protects small fish from being pulled into the pump.

Use this valve to control the water flow between the fountain and filter.

Main outflow to filter and/or waterfall.

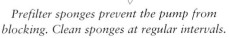

Prefilter sponges prevent the pump from blocking. Clean sponges at regular intervals.

Electric motor in a waterproof casing.

A self-cleaning pond pump filter

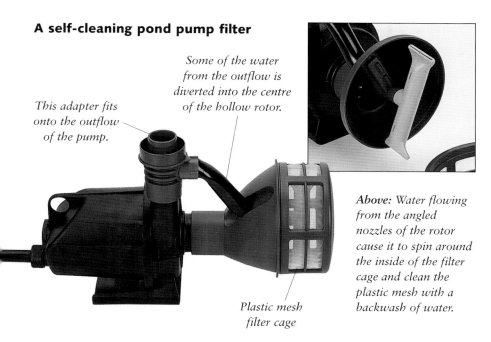

This adapter fits onto the outflow of the pump.

Some of the water from the outflow is diverted into the centre of the hollow rotor.

Above: *Water flowing from the angled nozzles of the rotor cause it to spin around the inside of the filter cage and clean the plastic mesh with a backwash of water.*

Plastic mesh filter cage

Choosing a pump

Choose a pump and filter with the correct flow rate and capacity for your pond. Ask the advice of a specialist water garden or aquatic outlet. As a rule, the volume of your pond should be circulated through the filter at least every two hours.

Be sure to power the filter with a good-quality, reliable pump. The pump is the heart of your pond and it is false economy to opt for a cheap, unreliable model that may break down in a short time.

When buying a pump, also pay particular attention to the flow rate. Pumps are generally sized at 'zero head', which means that the stated maximum flow rate is available if the pump does not have to lift the water any height. The higher the water has to be pumped, the lower the output

will be. For example, a pump with a maximum flow of 9000 litres per hour, may only pump 4500 litres if the water is lifted to a height of two metres, and may not pump at all at a height of four metres.

It is therefore essential to calculate how high your pump will have to push the water to supply the filter. If the filter requires a flow of 6750 litres per hour, make sure that your pump can achieve this at the height of the filter above the surface of the pond. If a fountain is being run off the same pump, allow one third of the flow rate for this function.

Remember that pump flow rates are generally calculated on the assumption that the water is clean and they do not take into account the clogging of prefilters. The maximum flow rate of many pumps

Selecting the best pump for your garden pond

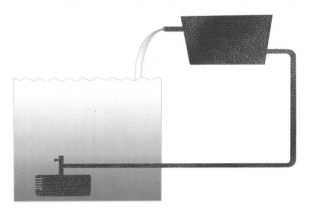

An external filter
When using an external box filter, select a good-quality pump powerful enough to push the volume of the pond water through the filter about every two hours.

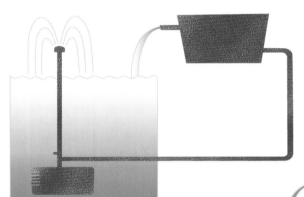

Filter and fountain
If you want to use the same filter but with a fountain playing in the pond, fit a bigger pump to allow an extra third of the flow rate to produce the fountain.

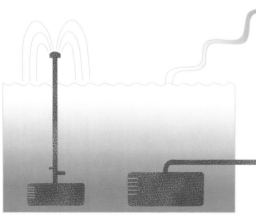

Waterfall and fountain
Having a fountain and raising the filter up to create a waterfall may demand too much from one pump. Use two pumps – one powerful enough for each task.

can be reduced by up to one third when used in a dirty pond.

When using your pump for a waterfall, remember that every 15cm (6in) width of a waterfall will require 1350 litres per hour to produce a thin sheet of water. Again, this flow is required at the height of the waterfall. A waterfall 45cm (18in) wide will therefore require a pump capable of delivering 4050 litres per hour. If the waterfall is 1.5m (60in) high, the pump must be capable of delivering 4050 litres per hour at 1.5m (60in).

Fortunately, most pump manufacturers supply charts showing the performance of their pumps at a variety of heights. If in doubt, buy a pump that is slightly oversized, as it is simple to restrict the flow with a variety of fittings.

When selecting a pump, take time to ascertain whether spares are readily available, and how long items such as bearings and impellers are

likely to last. A pump that appears good value at first may require regular spare parts and will therefore become expensive to run over time.

Pumps for different purposes

Pond pumps perform many functions. They may be used with an internal or external filter, a waterfall, fountain or feature ornament, or with any combination of all these. Different pumps are designed for different applications. Given the variety of parameters that a pond pump is required to meet satisfactorily, it is often a good idea to buy two pumps. One pump can be used exclusively for a fountain and/or ornament. The second – usually the larger – pump can power a filter, possibly using the outlet of the filter to provide the flow for a waterfall. Using two pumps will ensure that each application is correctly catered for. For example, a prefilter is extremely important for

Left: This pump being lowered into the pond supplies water to an external box filter at the top of a waterfall feature. Choose a pump with sufficient power to move water uphill to the required height and use another smaller pump to supply a fountain head or pond ornament.

a pump used with a fountain or ornament. Lack of a good prefilter will cause the fountain head or ornament outlet to block quickly, which means frustratingly frequent cleaning. Pumps being used for fountains and water features are not usually required to pump large volumes of water at significant heights and are therefore often less expensive to buy and run. They can be turned off at night, thus reducing overall running costs of the pond.

The 'main' pump, however, will be required to work with a filter and possibly a waterfall, too. Ideally, it

Above: A newly planted waterfall feature with an external box filter supplied from a pump in the pond. To maintain the flow, use the biggest bore pipes that will fit the pump outlet and filter inlet.

Right: A small pond with a fountain and water jet returned though a garden ornament. One pump with a foam insert is fine for this. Ensure that the electrical connections are properly installed.

will be able to pump small solids of 3-4mm (0.12-0.16in) without clogging. Filter pumps either have no prefilter or they have an impeller that will not block if the prefilter is removed. Removing the need for a sponge prefilter will ensure that the pump flow is not affected by the pre-filter clogging up. Dirty water from the pond therefore reaches the filter more quickly and ensures that the pond is cleaner, as all the waste is being taken from the pond and trapped by the filter media.

Electricity in the water garden

Electricity and water can be a dangerous combination! It is vital that you observe the correct safety precautions when connecting your pond appliances to an electricity source. Do not use extension cables or the plugs found on most domestic appliances. Aquatic retailers and garden centres stock a range of weatherproof junction, socket and sswitchboxes specifically designed for outdoor use. You can be sure that their fuses are correctly rated and the units are built to a sufficient quality. The most popular boxes available are switchboxes with two or three switches for controlling pond pumps, lighting and filtration.

Although there are economy versions available with only one fuse, it is always worth selecting a switchbox with individual fuses for each of the outlets. Individual fusing ensures that even if one appliance trips out, the remaining ones will continue to work. This can prevent fish losses due to unnecessary pump and filter stoppages.

Do not leave a switchbox or plug and socket lying around the garden where they can be damaged. Instead, fix them securely to a wall or strong post. Switchboxes are generally weatherproof but not waterproof; this means that they cannot be submerged in your pond.

COPING WITH ALGAE

Many pondkeepers experience problems with green water and blanketweed, both of which are generally harmless to fish, but are an eyesore. However, there are several methods – natural, mechanical and chemical – of controlling them.

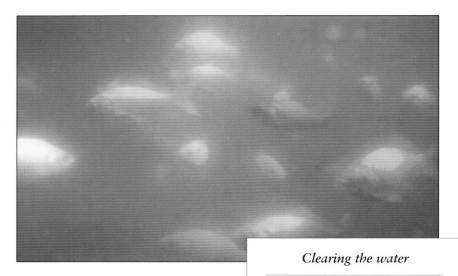

Clearing the water

Green water not only looks unsightly, but may also be a sign of poor water quality. Fortunately, there are several simple ways to remedy green water.

Algae thrive in bright sunlight, in a pond high in nutrients. It is, therefore, generally a summer problem, made worse by long hours of sunshine and high temperatures. However, aquatic plants are higher in nature's chain than single-celled algae, so as the plants begin to flourish and take up the excess nutrients, algae problems tend to decrease. Once plants begin to grow, and waterlilies and other surface-growing plants provide an adequate amount of shade, the algae no longer obtain the high levels of sunlight and nutrients they require and begin to disappear. Adding more aquatic plants, particularly the submerged varieties, can help to reduce levels of

Right: High algae levels can cause lack of oxygenation, as the surface of the pond becomes covered in a coating of algae. Remedy the situation promptly.

algae, and it is generally ponds with insufficient plant growth that experience algae problems.

The principal algae in the pond

Many species of algae can cause green water and unsightly, stringy growths in the pond. These species can be divided into two basic groups: those that cause pond water to appear green, with the consistency of pea soup, and those that form blanketweed, a carpetlike algae that covers rocks and plants and clogs pumps and filters.

The first type of algae is a single-celled plant that causes algal blooms in ponds. The pea soup effect is very frustrating and common in new ponds. There are many causes, but the most common is an excess of nutrients in the water, coupled with a lack of plant cover on the surface of

the pond. Excess nutrients can be the result of overfeeding, overstocking with fish or a lack of filtration. In a pond where everything is balanced, with the correct level of fish and plants, pond water should be clear.

Where new ponds are concerned patience is the key, because it takes time for a natural balance of nutrients to establish. Especially in the first season, it will be a while before plants put on good growth and there may not be much surface shade to reduce algal growth.

Controlling algae

It is important to avoid the urge to empty a pond full of green water. Although this may seem the sensible thing to do, nothing could be further from the truth. Emptying the pond will temporarily alleviate the green water situation, but minerals in tap

water will cause the refilled pond to become thick with algae very quickly. These salts and minerals are one of the causes of algae and take a long time to break down. If you are looking for a 'quick fix' for green water (not in a new pond), use one of the many proprietary algae treatments available from garden centres and aquatic shops.

There are several artificial ways of removing algae from the pond. In the case of ponds where you are unlikely to achieve a natural balance, such as koi ponds where there is typically very little plant life, it may be necessary to install permanent anti-algae measures, such as ultraviolet (UV) sterilizers. In other cases, where a temporary cure is required, you can add a liquid treatment.

Using chemical treatments
Pond algicides are available in various forms: most are liquid treatments, but slow-release algicide blocks and anti-algae tablets are also available. In theory, chemicals affect algae and leave other plants unharmed. They act by destroying

Right: Be sure to calculate the volume of the pond accurately before adding an algicide. To protect fish and plants from the risk of overdosing, always mix chemical treatments in a bucket or watering can and distribute them evenly around the pond. Do not exceed the doses recommended by the manufacturers.

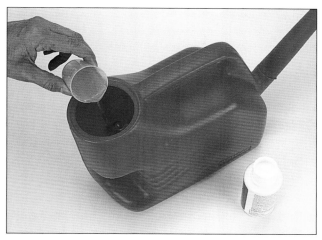

Left: Good plant growth will result in lower nitrate levels, thus reducing the food source for algae. A healthy pond should have a mixture of plant types.

the thin wall cells of algae, while leaving more robust plants unharmed. However, if over-used, algae treatments can restrict the growth of aquatic plants, so be sure to follow the recommended dosages. Do not use algicides in a new pond as they may restrict the growth of new plants, which are particularly susceptible before they have established themselves in a pond.

Algae treatments should always be considered as short-term measures. For a long-term solution, look at balancing the nutrient level in the pond with correct planting, filtration and stocking of fish.

When using algae treatments it is important to monitor water quality, as there is a danger of deoxygenation caused by the algae dying off and decaying. Before adding algae treatments, change 10-20% of the water to remove a proportion of the dead algae and reduce the likelihood of pollution of the pond. As an alternative strategy, use a very fine foam or pad in the pond filter to remove small algae particles.

Ultraviolet (UV) sterilizers

Ultraviolet sterilizers, or clarifiers as they are also known, work by passing pond water close to an ultraviolet lamp that breaks down algae cells. Many units are provided with a 'clear water' guarantee. Ultraviolet sterilizers are extremely effective when used correctly and are the most popular form of long-term algae preventative.

Water passing through the UV unit should be as clean as possible and will ideally have been passed through a filter first. Prefiltering the water before subjecting it to UV light achieves two things. First, the algae are less likely to be obscured from the UV light by clumps of particulate and plant matter. The 'kill rate' will

A conventional UV sterilizer

This unit is designed for use outside the pond.

Weatherproof housing for safe outdoor use.

Quartz sleeve protects lamp and allows UV penetration.

Linear UV lamp treats circulating water.

UV resistant housing provides robust protection for quartz sleeve and lamp.

Transparent hose tails glow blue when the UV lamp is working.

An underwater UV sterilizer

This UV unit is designed for use inside or outside the pond.

*Treated water leaves
the sterilizer here.*

*Water is
pumped in
here.*

*Water is treated as it circulates around a 'PL'
ultraviolet lamp housed in a sleeve made
from a derivative of ptfe, a type of plastic.*

therefore be far higher than if the UV has to penetrate less clear water. Secondly, non-filtered water may cause the UV to clog and will greatly increase the time you have to spend on maintenance.

Choosing the right UV sterilizer

UV filters are rated according to the wattage of the lamp being used. When you buy a UV unit, make sure you buy the correct sized unit for your pond. It is not only the lamp size that is important; pay attention also to the flow rate of pond water through the unit. If it is too slow, the unit will not recirculate water at a sufficient speed to function efficiently. If it is too fast, the water will not be in contact with the UV lamp for sufficient time to kill the algae. Always seek advice from your local aquatic centre regarding the correct size and flow rate for your UV sterilizer.

There are different types of UV lamp available today. The most popular ones are linear, with connecting pins at both ends. These lamps are relatively cheap to replace and have an effective life of six to nine months. They are housed in a quartz sleeve to protect them from contact with water. The quartz allows the UV light to penetrate the water and kill the algae cells. The second type of lamp commonly available is the 'PL' lamp. It is more expensive to buy, but has a longer effective life in the sterilizer.

The effectiveness of ultraviolet sterilizers is largely dependent on maintaining the units correctly. Be sure to replace the lamps regularly,

COPING WITH ALGAE

following the manufacturer's recommendations. It is also vital to keep the UV units clean. The higher-quality quartz sleeves available are more resistant to 'fogging' than softer glass, but regularly inspecting and cleaning the quartz sleeve will ensure that the sterilizer operates more effectively. If the glass has become permanently discoloured, replace it. When handling quartz sleeves, be careful not to leave fingerprints on the glass, as grease will prevent light from penetrating the glass efficiently.

Always handle quartz sleeves carefully, as they are extremely brittle and will crack if knocked.

Recent advances in ultraviolet sterilization have led to the development and availability of submersible units. These units may be used both in and out of the water, and give you the opportunity to hide a potentially unsightly piece of equipment.

The latest lamp technology has resulted in sterilizers with non-quartz sleeves. These new sleeves are made from a derivative of ptfe, a plastic material used on 'non-stick' cookware, and are far more durable, less resistant to fouling and much easier to clean. Although these units may be slightly more expensive, they are worth considering because of their significant advantages over more traditional sterilizers.

During the winter months, remove UV sterilizers from the pond for cleaning and install replacement lamps ready for the following spring.

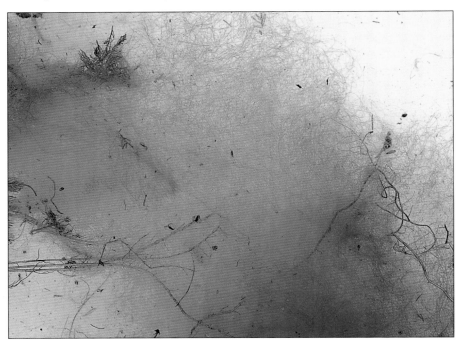

Above: Before using any chemical remedy against blanketweed, use a stick to remove large algae growths. This helps prevent water quality problems caused by blanketweed dying in the pond.

Left: Blanketweed will grow quickly if given the opportunity and will rapidly smother slow-growing plants in the pond. This close up shows the long filaments that make up the dense green masses.

This is also a good opportunity to ensure that the electrical end caps are in good condition.

UV sterilizers must be used with care. Never look at the burning lamp and ensure that electrical connections are made safe.

Controlling blanketweed

Blanketweed, thread algae, silkweed and filament algae are all different names for basically the same type of algae. Most commonly know as blanketweed, this algae is made up of many different species of filamentous algae. Blanketweed is often seen clogging pond filters and tangling itself around pond plants, notably oxygenating plants such as *Elodea crispa* and *E. canadensis*.

In a balanced pond, although there may be signs of blanketweed, it rarely becomes a problem. However, as most pondkeepers struggle to reach a natural balance, or make no effort to achieve it at all (as in koi ponds), blanketweed can become an unsightly nuisance.

Blanketweed can thrive in many different conditions, but does best in well-lit ponds that are high in nutrients. Ponds with little or no plant life are often plagued by blanketweed problems. Not only is there no protection from sunlight, which enables the blanketweed to prosper, but there are no plants to take up nutrients from the tap water used to fill the pond and the food used to feed the fish.

Although filtering a pond will help to control free-floating algae, it will not aid the removal of blanketweed. Indeed, the increased use of filters and ultraviolet sterilizers raises the chances of a blanketweed problem in a pond, as these filters will provide clear water, high in nutrients through lack of competition from other forms of algae.

The occurrence of blanketweed has increased greatly since more pondkeepers began using ultraviolet sterilizers. The lack of suspended algae in the pond ensures that there are nutrients for blanketweed to

thrive on, and sunlight is able to penetrate the water, again aiding in blanketweed growth. Once it is growing in a pond, blanketweed quickly spreads and soon becomes an eyesore.

There are many methods of tackling blanketweed and most meet with some success, at least for a while. However, it does appear that this form of algae becomes resilient to certain forms of control over time. In essence, the blanketweed appears to adapt in order to thrive against the treatments being used against it.

Below: Barley straw is a traditional method of reducing blanketweed. Algae control mats are now available and can provide a more effective solution.

These pads are impregnated with an algicide.

These mesh bags contain barley straw. As they decompose, microbes help to disperse algae. Simply place them in the pond.

Place them in an external box filter or weight them down in the pond.

Renew them regularly.

Eliminating blanketweed

Algicides are the easiest and quickest method of trying to remove blanketweed. They vary from simple light inhibitors to complicated chemical formulae designed to kill the blanketweed. Be careful when using algicides and follow the recommended dosages. Be sure to remove as much blanketweed as possible before treating the pond. As the algae dies off, remove it as quickly as possible. Large growths of blanketweed dying off in the pond can cause water quality problems, so monitor the ammonia and nitrite levels while using any method of killing blanketweed.

Barley or lavender straw added to the pond or filter in net bags is a natural method of blanketweed removal that sometimes works, but does not provide consistent results. Do not allow these products to decompose in the pond and replace them regularly.

Vegetable filters are used in a small number of ponds. Plants such as watercress are good nitrate removers. The drawback is that they need to cover a large surface area, even in a relatively small pond.

Pond algae magnets have become popular in recent years. They work by using their polarities to break up the calcium carbonate in the water, thus weakening the thin-walled cells of blanketweed. Algae magnets work best when used intermittently. Try using a magnet for two to three weeks, then removing it for the same period before reapplying it. Prolonged use appears to allow the blanketweed to become resilient to its effects. In fact there are several types of blanketweed and it is more likely that a new type takes over when the initial form has been removed from the pond water.

Below: Algae magnets are an aid to blanketweed control. The strong magnetic fields disrupt the algal cell structure. Best used intermittently to prevent new forms taking over from disrupted ones.

An electronic blanketweed controller *This unit is for use outside the pond.*

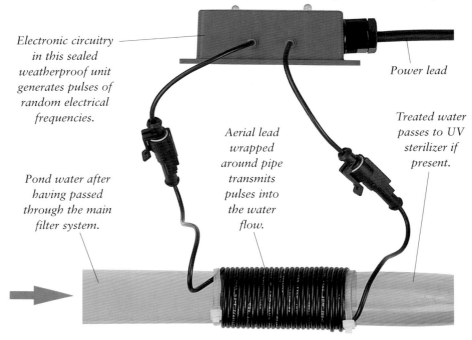

Electronic circuitry in this sealed weatherproof unit generates pulses of random electrical frequencies.

Power lead

Treated water passes to UV sterilizer if present.

Aerial lead wrapped around pipe transmits pulses into the water flow.

Pond water after having passed through the main filter system.

Electronic units are the latest and most successful form of blanketweed remover. There are two basic types. The cheaper units replicate the action of an algae magnet by supplying a static magnetic force to the pond water. Because the magnetic field is not constantly changing, use these units in the same way as magnets by running them intermittently. The more effective units use a small processor that alters the magnetic field so that the blanketweed is not allowed to adapt to the constantly changing conditions.

These units work by coiling their aerial wires around the filtration pipework. The aerials provide a continually shifting polarity that weakens and disturbs the calcium ions in the pond and causes the blanketweed's thin cells to break down. Electronic blanketweed controllers are safe for use with all other plants, which have far stronger cell structures than blanketweed.

All the methods of controlling blanketweed described here will achieve varying degrees of success. However, with good water management and sensible fish and plant stocking levels, algae and blanketweed problems will be kept to a minimum by nature itself.

Right: The correct combination of plants, fish and filtration will result in a healthy pond with clear water all year long.

GROWING HEALTHY PLANTS

The regular care and maintenance of plants is important in the establishment of a balanced and visually attractive water garden. Once established, aquatic plants grow rapidly and will require dividing and cutting back to look their best.

Keeping plants in trim

Pond plants play a vital role in the healthy pond. Regular manicuring ensures the maintenance of a mature appearance without overcrowding.

In order to secure a good display in the first season, it is best to plant pond plants during late spring and early summer. Later plantings establish without difficulty, but they have to be cut back hard before planting and this affects their early summer appearance.

Choosing healthy plants

Freshly bought or divided plants can be planted without much preparation in early spring, but inspect them carefully for small cylinders of jelly containing the eggs of the greater pond snail, which may be clinging to the stems and foliage. Be sure to remove these, as the resultant snails are an annoying pest that can severely damage plant foliage. All plants for planting should be well-balanced, the top being compatible in size with the root. Do not be worried about removing foliage from any

Right: Before adding new plants to the pond, examine them carefully for evidence of the cylinders of jelly containing the eggs of the troublesome greater pond snail. They are often mistaken for fish spawn, although they bear little resemblance to one another.

well-rooted aquatic plant. As long as the plant is in good health it will grow away again very quickly.

Preparing and planting waterlilies

When you buy bare-rooted water-lilies, they often have some leafy growth. If the leaves are allowed to remain after planting, they usually die. They may also act as an aid to buoyancy, lifting the plant right out of the container into which it has been freshly planted. So cut the foliage right back to the crown, leaving only the spearlike, submerged foliage. Also trim the roots and pare back any signs of decay on the rootstock to sound material. Existing roots, like the foliage, will almost certainly die following transplanting and are best cut back to allow fresh ones to develop.

Theoretically, waterlilies that are already established in containers when you buy them can be placed in the pond without them realising that they have been moved. The reality is rather different, for even if the plants are propped up on bricks to ensure

that they are at exactly the same level in the water as they were in the sales tank, the leaves are likely to curl and no longer lie flat on the water. It is much better to remove all floating foliage and to position the plant at the required depth.

Preparing and planting marginals

Tall-growing marginal plants must also be cut back before planting. Often by mid-season the foliage is completely out of proportion to the root system. Even young plants already established in containers, which just require placing in the pond, can be encouraged to grow more strongly and thickly if you first cut them back.

Preparing submerged plants

With submerged plants, the more vigorous the young shoots the better. When preparing new bunches of cuttings from existing plants, remove a number of sprigs of healthy growth and wrap a thin strip of lead around the base. When planting, make sure that the lead is completely buried, otherwise it will rot through the stems and the cuttings will float to the surface of the pond.

Growing plants in containers

In the majority of ponds, cultivating aquatic plants in containers makes for easy maintenance and management. Individual varieties can be lifted and divided as required and isolating one kind from another means that they do not readily invade one another's territory.

Open latticework-sided containers are best for most aquatic plants, as they permit the healthy exchange of gases that would be restricted by a solid pot. Waterlilies and other deep water aquatics can be established within a closed pot or container, but will start to deteriorate after about 18 months. The growing medium often turns a blackish-blue and smells unpleasant. Submerged aquatics are the only rooting plants that grow successfully in closed containers.

Modern baskets are manufactured from tough plastic and are generally tapered to assist with the balance of the plants. In high winds, tall container-grown reeds or rushes can easily topple over. Plants that are regularly retrieved from the water after a breeze always look forlorn and never develop satisfactorily.

At one time, it was essential to line aquatic planting baskets with a square of hessian to prevent the potting mixture spilling into the

Planting *Caltha palustris* in a flexible container

1 This flexible container is easier to wedge into tight corners than rigid baskets. Part-fill it with soil, make a hole, spread the roots and firm in the plant.

2 Water freely to drive all the air out of the soil. This prevents any air bubbling out and possible escape of soil when the container goes into the pool.

water. However, with modern micromesh containers this is unnecessary. Loosely fill the container to the top with heavy soil or an aquatic planting mixture and then firmly plant the aquatic, rather as if it were being put into a conventional pot. Take a watering can with a fine rose and thoroughly soak the planted basket. This drives much of the air out of the planting mix and prevents vigorous bubbling and the possible dirtying of the water when the container is lowered into the pool.

Thoroughly watering the basket causes the planting mix to settle. Top it up, firm it by hand and water it again heavily until most of the air is driven out and it is just below the surface of the basket. Top off with fine pea gravel and water again thoroughly. The gravel will prevent fish from stirring the soil when searching out aquatic insect life. Once the baskets have been prepared, place them in the pond.

For almost all ponds, it is essential to plant in containers, even though it may seem more natural to plant directly into soil spread across the pool floor. Obviously this is how plants would grow naturally, but from a management point of view it is a nightmare, as many pond plants are vigorous-growing and quite invasive. If unrestricted, the faster-growing kinds rapidly overtake the more moderate species and a pond can lose up to half its plant diversity within three years, with the stronger

3 *Place well-washed pea gravel around the plant, covering the entire surface. This helps to prevent fish from stirring up the soil and clouding the pond water.*

4 *Soak the planted container again before placing it in the pond. All the air should have been excluded and any dust on the gravel washed out.*

Planting myosotis in a rigid plastic basket

1 Remove the plant carefully from its pot and tease out the roots if the rootball is congested. Make a suitably sized hole in the soil and place the plant in the centre of the basket.

2 Use a trowel to place more soil around the plant and firm it in, leaving sufficient room for a topdressing of well-washed pea gravel. The soil level will drop when you water the basket.

3 Water thoroughly and then distribute pea gravel over the soil surface to prevent fish from stirring it up and dirtying the water. The gravel also adds a finishing touch to the planted basket.

4 Once planted, thoroughly soak the container again using a watering can with a fine rose. This helps to prevent disturbance when the planted basket is lowered into the pond.

kinds smothering the weaker-growing varieties of pond plants.

The other problem, if you choose to establish aquatic plants directly into the pond, is coping with cleaning out when this is necessary. In a natural earth-bottomed or concrete pond this is not going to present any serious difficulties, but it is a very different matter for a pond made with a liner, where every tug on a plant or movement of the spade has the potential to cause a leak. Growing aquatic plants in baskets that can be lifted and removed for maintenance overcomes this problem.

General care and maintenance

If a garden pond is to be a constant source of pleasure and delight, it will require a certain amount of maintenance. Immediately the winter has passed, remove any lingering plant remains and mulch the bog garden with well-composted bark or peat. Mulching with organic matter every year helps to improve the soil conditions, but the mulch must not

be rich in nutrients, otherwise they will leach into the pool and cause the development of an algal bloom.

While aquatic plants do not benefit from a traditional mulch, it is useful to go round all the containers and topdress them with pea gravel. At the same time, you can inspect the plants to see whether they are ready for division and remove any stray shoots that are emerging erratically from the baskets.

If the pond needs cleaning out, do this in the spring. It need not be an annual event; you can usually leave a well-ordered pool untouched for seven years or more. When it does prove necessary, make arrangements to keep the plant containers wet. Only submerged aquatics and waterlilies need be kept under water while the cleaning out takes place.

Apart from fertilizing the plants in summer, it is usually necessary to thin out weak shoots and remove fading foliage and blossoms before they have an opportunity to set seed. This is very important with plants

Left: In the autumn it is important to remove all organic debris, such as fallen leaves and decaying plant remains. This will decompose in the water and may cause problems for the fish.

Above: Floating aquatics, such as fairy moss (azolla), can become a nuisance if not regularly netted off the pool. However, properly controlled, they make an important contribution to pool clarity.

summer is a big mistake. It is regular manicuring of the plants that makes all the difference to the appearance of the pond.

The same principle applies to floating plants, especially the potentially invasive azolla, or fairy moss. This is an invaluable plant if you can control it, as it makes a major contribution to ensuring clear water. However, if it is netted off the pool to excess, the clarity of the water may be affected. Fairy moss often sticks to the surface of marginal containers and can look untidy, so float it off from time to time by raising the water level.

As part of your routine maintenance, you will need to replace all aquatic plants on a regular basis, usually by division or cuttings. Not all have the same rate of growth. There are differences even amongst related plant groups, so it is difficult to provide general instructions. Waterlilies of the more vigorous varieties must be divided every three or four years, while some of the smaller-growing and pygmy kinds will last for five or six years if regularly fed and topdressed.

A preponderance of smaller than typical leaves, together with irregular flowering, is an indication that a waterlily or other deep water aquatic is in need of lifting and dividing. In the case of waterlilies, particularly the stronger-growing varieties, this uncharacteristic growth is often accompanied by central clumps of leaves that climb out of the water. If this happens, divide the plants quickly, although if you see it from

such as the water plantain, because not only is seed production a drain on the plant, but if viable seed is produced and scattered, the emerging seedlings can create a troublesome weed problem. This can happen with some of the reeds and rushes, too.

Carefully examine submerged aquatics. Providing that a balance has been struck and the water in the pool is clear, regularly remove any messy and unattractive surplus growth, together with any woody material. However, a mass clearout of submerged aquatics during the

GROWING HEALTHY PLANTS

late summer onwards, it is better to wait until the following spring before undertaking the task.

Dividing hardy waterlilies

Spring is the ideal time to divide hardy waterlilies. Lift and wash each plant and remove any adult foliage. You will see that each plant consists of a main rootstock from which several eyes or shoots have grown to form sizable branches. Retain these side growths, cutting them from the parent plant with as much rootstock as possible. The thick, bulky part of the original plant is generally of little use and should be discarded, but all

Below: When waterlilies become crowded, they produce small flowers and the leaves push out of the water. This is a sure sign that it is time to lift and divide them. Divide hardy waterlilies in spring.

the branches can be planted to form new plants if each has a healthy terminal shoot.

The majority of waterlilies can also be increased in the spring from eyes. These are tiny growing points that occur with varying frequency along the rootstocks. They look like small versions of the main growing point, each with their cluster of leaves, although with a few varieties they appear as brittle, rounded nodules that are easy to detach.

Remove the eyes with a sharp knife, and dust the exposed tissue of both eyes and the rootstock with powdered charcoal to protect against infection. Plant individual eyes into pots of aquatic planting mixture or good, clean, heavy loam and stand the pots in a shallow container with enough water to cover their rims. As the young shoots develop, raise the

Propagating waterlilies from eyes

1 Remove the eyes or small shoots from the main rootstock with a sharp knife. Be sure to remove part of the starchy rootstock with the shoot.

2 Trim each eye of superfluous roots or debris. Wash thoroughly. If the cut surface looks vulnerable to infection, dust the wound with powdered charcoal.

3 Place each eye into an individual small pot of aquatic planting medium. Firm in well, leaving just the tip of the shoot above the soil surface. Water thoroughly.

4 Place each pot into a bowl of tap water, making sure that the surface of the medium is completely submerged. As the leaves develop, add more water.

water level and transfer the plants into successively larger pots until they are big enough to be installed in a large planting basket.

Maintaining marginal plants

Marginal plants should be regarded rather like those in the herbaceous border and lifted as necessary. There are great differences between the rates of growth of, say, a rampant reedmace and the more measured development of a skunk cabbage. Once a plant is identified as being overcrowded, remove it from the pond and divide it, ideally during the spring. This gives it the opportunity to develop into a reasonably mature and attractive plant during the first growing season. Dividing up a plant once growth is under way not only results in a less attractive specimen,

but often impairs flowering as well. Irises are the only plants to benefit from dividing during the summer. If they are disturbed during the spring, then summer flowering will be greatly affected. Lifting and dividing irises immediately after they have flowered, gives them an opportunity to become well established for the following season and then make a fine display. Do not be nervous about cutting the foliage back hard and separating the clumps into individual fans of leaves. Be sure to trim the roots back sharply before replanting.

For the majority of aquatic plants, division is simply the process by which a crown is separated into sustainable portions that are potted or planted out individually. In the case of plants such as reedmaces and rushes, which have creeping root

Increasing butomus from turions

1 This mature butomus has a strong, creeping rootstock and dense clusters of turions. Some of those from the previous year are developing into new plants; separate them and trim the roots.

2 Place three or four turions in a small pot of aquatic planting mix. Firm in gently. Grow on until sufficiently large to pot individually. Stand the turions in water just covering the rim of the pots.

systems, this often involves removing lengths of rootstock, each with a terminal bud, trimming these up and potting them individually. Effectively, each division is a large bud or shoot with a small cluster of roots.

Dividing bog garden, submerged and floating plants

Most aquatic and bog plants are usually divided during early spring, just as their shoots are emerging. However, those that flower at that time, such as marsh marigolds and the early drumstick primulas, are best dealt with as soon as their blossoms have faded. Astilbes, hostas and other popular bog garden plants are sometimes divided during the depths of winter, and while this is perfectly satisfactory, it is often easier to undertake in the spring. In the autumn or winter it is often necessary to use a knife to separate the plants, while in the spring, once the plants have started into growth, they tend just to pull apart.

Several aquatic plants produce winter storage organs, or turions. In some cases, as with the arrowhead (*Sagittaria sagittifolia)* or flowering rush *(Butomus umbellatus)*, these are primarily food storage organs. However, they are produced freely and if you separate them carefully, you can establish them as individual plants. If left to their own devices they form sizable solid clumps, which in due course have to be lifted and separated if the plants are to continue to develop satisfactorily. With some submerged plants and a number of floating aquatics, the

production of winter turions is a definite effort to multiply. This especially applies to the frogbit *(Hydrocharis morsus-ranae)*. As autumn approaches, it breaks up into a dozen or more individual plantlets, initially attached by runners. Eventually they turn into hard, fat, leafless buds that sink to the floor of the pool in order to overwinter.

The water soldier *(Stratiotes aloides)* behaves in a similar way, but usually retains an old brittle plant with tiny plantlets clustered round it, each attached by a runner, rather like the indoor spider plant. In due course these separate or you can divide and redistribute them in the spring to raise new plants.

Feeding plants

Each season there will be some plants in the pond that do not require lifting, dividing and replanting. However, most certainly benefit from feeding. There are two common methods of providing fertilizer for established plants without polluting the water. The first is a traditional and now questioned method, the other is straightforward and easy to administer.

The traditional method involves a mixture of clay and coarse bonemeal. These are combined to create 'bonemeal pills', which are pushed into the planting medium next to each plant. Each pill consists of a handful of coarse bonemeal with sufficient wet clay to bind it together. It is a very practical way of introducing a slow-release fertilizer into an aquatic environment without

Dividing irises

1 Divide irises after flowering. Wash the lifted clump well and separate out individual fans of leaves. They should pull apart quite easily. Select the younger, more vigorous individuals for replanting.

2 Using a sharp knife, cut away the bulk of the fibrous roots from separated plants, as this is likely to die back anyway. Shorten the foliage so that the plant is well-balanced. The newly potted plants will then establish quickly.

The gravel will prevent fishes from disturbing the newly planted iris.

3 Plant the divisions singly into small mesh containers filled with aquatic planting mixture. Water thoroughly and cover the surface with a layer of rounded gravel. Place the basket on the marginal shelf of the pond, so that the water comes over the top of the basket.

Aquatic soil and fertilizers

You can buy specially formulated soil for use in aquatic planting baskets.

Peeling off the label on these plastic sachets exposes two holes that allow the fertilizer to leach slowly into the soil.

These are slow-release feed tablets. Simply bury one or two in the soil close to the plant roots.

polluting the water. The only uncertainty is the speed of availability of nutrients to the plants, as recent research has suggested that bonemeal takes very much longer to release plant foods than had been previously believed.

The modern method of fertilizing aquatic plants is to introduce specially manufactured aquatic fertilizer tablets or sachets of slow-release fertilizer to the planted baskets. The sachets usually consist of a small package with a perforated section through which the fertilizer is slowly released. Liquid fertilizers for pond plants are also available.

Propagating from stem cuttings

While it is not necessary to plan a propagation programme for a pond in the way that you might for a bedding plant scheme or a vegetable plot, the ability to rejuvenate ailing

aquatic plants from seed or cuttings is very desirable. The majority of aquatic pond plants are easy to propagate this way.

Several marginal plants, many submerged aquatics and a few bog garden plants benefit from regular regeneration from stem cuttings during spring and early summer. In the case of marginal aquatics, such as the water mint *(Mentha aquatica)* and brooklime *(Veronica beccabunga)*, take cuttings from overwintered parent plants in spring, once the shoots are in active growth. Take the cuttings as early as possible in the season to replace untidy and woody parent plants. Elderly water mint and brooklime rarely flourish in the same way as freshly rooted cuttings treated as annuals.

Remove shoots up to 5cm (2in) long, cutting at a leaf joint, as it is in the area of the leaf joint that the cells that will be stimulated into producing roots will be most active. Remove the lower leaves so that they do not rot off in the water. There is no need to use a hormone rooting preparation, as most aquatics root quickly in very wet mixture or mud.

Put the pans or trays of cuttings in a container with water, just covering them. Shade the cuttings from direct sunlight and within 10-14 days they should be rooted. Then pot them on individually and allow them to produce small tight rootballs before planting several together in an aquatic planting basket. If the cuttings start to grow vigorously without branching, pinch out the top to encourage bushy growth. With the exception of hair grass *(Eleocharis acicularis)*, which is increased by division, all the other popular submerged aquatics are increased or replaced by stem cuttings. The majority should be propagated from vigorous young growths early in the season. Later on the shoots become rather brittle.

Remove good, healthy cuttings of non-flowering shoots, tidy up the foliage and gather together small bunches of 5-10cm (2-4in)-long

Taking cuttings of *Mimulus ringens*

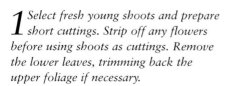

1 Select fresh young shoots and prepare short cuttings. Strip off any flowers before using shoots as cuttings. Remove the lower leaves, trimming back the upper foliage if necessary.

2 Fill a 7.5cm (3in) plastic pot with aquatic mix. Make it wet. Arrange four cuttings around the edge and submerge the pot in water. Keep cuttings in shade; they root in about 10-14 days.

stems. Fasten them neatly around the base of the stems with a thin strip of lead. This not only holds the bunches of cuttings together, but also weighs them down, an advantage when they are planted in their permanent container. Plant the bunches of cuttings in aquatic planting baskets filled with good-quality aquatic plant mixture or heavy loam or clay soil. Bury the lead weight beneath the potting mixture to prevent it rotting through the stems and topdress the baskets with pea gravel.

Above: The seeds of the majority of aquatic plants germinate more readily when sown directly after harvesting. Fresh iris seeds should emerge within three to four weeks of sowing.

Propagating from seed

A number of aquatic and bog plants can be raised from seed. This is one of the easiest propagation methods, but remember that it is the species rather than the named varieties that come true. However, amongst bog garden plants there are several named coloured strains of primulas and irises that can be increased this way. Certainly it is the best way of producing quantities of plants quickly and economically.

Sow the seeds of aquatic plants in a wet, heavy loam potting mixture that has been thoroughly soaked. If the plants are going to produce floating leaves, cover them with a few centimetres of water. Ideally, sow seed in plastic pots or pans of potting mixture and use a small plastic aquarium or bowl as a plant nursery. When sowing seeds, be sure to cover them with potting mix and a thin layer of silver sand. This prevents them from floating away and the silver sand acts as a deterrent to algae invading the surface of the mixture. Once the seeds have germinated and the seedlings are established, prick them out into seed trays full of aquatic soil and place them in shallow water.

Raise bog garden plants in damp conditions, rather than very wet or submerged ones. The seeds of most bog garden plants are best sown fresh, during mid- to late summer shortly after gathering. Packeted seed rarely becomes available until winter or early spring. In some cases, particularly with bog garden primulas, seeds that are not fresh

Left: Sow seeds of aquatic plants directly onto the surface of aquatic soil in a shallow container. Keep the soil moist and put the container near a window for a few weeks.

because if fish are living in the pool, it is difficult to use pesticides without putting them at risk. The answer is to understand the life cycle of each pest and disease in order to be able to destroy it at its most vulnerable stage.

Pests of water plants

Waterlily aphids are probably the most common of all the afflictions of aquatic plants. These breed at a tremendous rate during warm humid weather and attack the flowers and foliage of waterlilies and other succulent aquatics. In fact, they have much the same kind of effect on aquatic plants as black bean aphids have on broad beans.

Eggs from the late summer brood of adults are laid on plum and cherry trees during early autumn, where they overwinter. The eggs hatch in spring and winged female adults fly to the plants, where they reproduce freely. As autumn approaches, a winged generation is produced, which then flies to the plum or cherry trees and deposits eggs.

During the summer, spraying the foliage regularly with a strong jet of clear water is the only way to control these pests. As they are knocked into the water, the fish clear them up. Insecticides are not much use unless

require chilling in order to break their dormancy. When this is necessary, sow them in a good soil-less potting mixture and treat them as if they were fresh. Water them thoroughly, then place the seed pans or trays in the freezer and leave them there for a week or two before returning them to the light and warmth. Germination generally follows quite quickly. Once the seeds of bog garden plants have germinated, prick out the seedlings and when they are established, transfer them carefully to individual pots for growing on until they are large enough to be planted in their permanent positions.

Pests and diseases

Aquatic plants are no different from any others in that they suffer from a variety of pests and diseases. Some of these can be highly troublesome,

the pond is free from fish. However, much can be done during the winter to break the life cycle of the aphids by spraying nearby plum and cherry trees with a winter wash. This destroys overwintering eggs.

Waterlily beetles can be equally irritating. The distinctive black larvae with yellow bellies strip the waterlily leaves of their epidermal layer and they begin to disintegrate. The small dark brown beetles hibernate in poolside vegetation during the winter and migrate to the waterlilies during early summer, where they deposit

Below: Waterlily aphids feed on all succulent pond plants. Applying a winter wash to nearby plum and cherry trees kills the overwintering generation and breaks the life cycle of these pests.

their eggs on the leaf surfaces. After a week or so the larvae hatch out and begin their trail of destruction across the waterliliy leaves. There can be several broods in a season.

The only practical control method is to hose the leaves down regularly to dislodge the pests so that they can be cleared up by the fish. Removing marginal plant foliage during the winter reduces the opportunity for overwintering beetles to hide close to the pond. Once waterlily beetles become well established in an area, constant vigilance will be required to prevent them devastating the plants.

False leaf-mining midge can produce superficially similar symptoms to waterlily beetle, as this tiny creature has larvae that eat a narrow tracery of lines over the

surface of the foliage of floating-leaved aquatics. The damaged areas start to decay and the leaves become detached and start to decompose. Rarely is damage as extensive as that caused by waterlily beetles, but it can be very disfiguring. There is no cure except the regular forcible spraying of floating foliage with a clear jet of water. Occasionally, in severe cases, the floating leaves are best removed and destroyed.

Caddis flies are pests with larvae that cause considerable damage to a wide range of aquatics. The adult flies visit the pond in the evening, depositing eggs in a mass of jelly that swells up immediately it touches the water. When the eggs hatch out, the larvae immediately start to spin silky cases that surround them, collecting all manner of pond and plant debris. At the same time they feed freely on any aquatic plants. Eventually, they pupate in the pool or amongst marginal plants around the edge. Spraying or hand-picking controls are impossible and so a great dependency is placed upon fish.

The brown china mark moth larva also takes pieces out of the foliage of aquatic plants and provides a shelter for itself prior to pupation by sticking down leaf fragments in which it weaves a greyish, silky cocoon. The damage is sometimes extensive, plants having torn and distorted leaves that are subject to fungal attack and then start to crumble around the edges.

The eggs of this insignificant moth are laid during late summer in neat rows on the foliage of any floating-

Above: *The greater pond snail will feed on algae, but much prefers to graze on succulent aquatic plants, causing considerable damage. They may be introduced on plants added to the pond.*

leaved pond plant. The eggs quickly hatch and the tiny caterpillars emerge and burrow into the undersides of the succulent foliage, in due time making small oval cases out of these leaves. During the winter they hibernate, re-emerging in the spring to continue their trails of destruction.

Pick off small infestations by hand, netting off and discarding all pieces of floating foliage in case they

have cocoons attached. If damage is widespread, cut off all the floating foliage and remove it. This enables the plants to have a fresh start and regeneration is usually quite rapid.

Pond snails can be very troublesome, especially the greater pond snail. It is often introduced to a pond in the mistaken belief that it will feed on algae. It may graze on blanketweed, but much prefers the soft, succulent floating foliage of waterlilies. As a general rule, avoid pointed-shelled aquatic snails, as most will feed on pond plants. Picking undesirable snails out by hand is the only reliable method of control, although you can also float a fresh lettuce leaf on the water overnight. The snails will have congregated beneath it by morning and are easy to remove and discard.

Diseases of water plants

Few diseases attack aquatic plants and with the exception of waterlily crown rot they are fairly innocuous. The two common waterlily leaf spots are disfiguring, but rarely if ever cause serious problems. One causes dark patches to appear on the surface of floating foliage. These eventually rot through and, in severe cases, cause the leaves to disintegrate. The other starts along the edge of the leaves, causing them to turn brown and crumble. There is no set pattern

Left: Mildew often attacks marsh marigolds towards the end of the growing season. It is disfiguring rather than debilitating. Cut off the affected foliage or use a systemic fungicide in severe cases.

for attack; in some years the problem is persistent, in others it does not appear at all. Remove disfigured leaves as they occur.

Waterlily crown rot is the worst aquatic plant disease and is still not well understood. It is believed to have been imported on waterlily crowns from the Far East and attacks all waterlilies irrespective of variety, causing the crowns to collapse into a decomposing brown mass. As there is little understanding of the precise cause of this particular rot, it is difficult to recommend a strategy for its control, although at present commercial growers recommend cleaning out the pool thoroughly and sterilizing it with a solution of sodium hypochlorite before flushing it out with fresh clean water.

Occasionally, marginal aquatics, especially marsh marigolds or calthas, suffer from mildew. This is a greyish mould that attacks the foliage, usually long after flowering. It is ugly rather than debilitating and in most cases is best dealt with by cutting off infected foliage. If the mildew is causing real problems for the plant, as sometimes happens with the white flowered *Caltha palustris* var. *alba*, and it is established in a container, then it can be removed from the pond in its container and sprayed with a systemic fungicide. Modern systemic chemical treatments are inactivated when they reach the soil or potting mix and once they have dried on the foliage present no problems for fish or other aquatic creatures when the plants are reintroduced to the pond.

CHOOSING YOUR POND FISH

No matter what the size or style of your pond, make it complete with the moving splashes of colour that only fish provide. From the smallest minnows to the largest koi, there is a variety of sizes, shapes and colours to choose from.

To keep your fish healthy, you will need to understand something about the environment in which they will thrive, their requirements and how to provide a compatible mix of fish that will thrive in your pond.

The right fish for your pond
Choosing the right size and species of fish largely depends on the size of your pond and the quality of the filter system. Some fish require well-oxygenated water and large swimming spaces, while others are at ease in small ponds with little water

Fish – the vital ingredient

Fish add colour and life to the pond. There is a wide variety of shapes and colours available amongst fish suitable for a garden pond.

flow. It is all too easy to buy a fish that you find appealing, only for it to outgrow your pond or eat the other inhabitants. Once the pond is fully stocked, do not be tempted to add one more fish just because you find a brighter, bigger specimen. Allow the pond to flourish with the existing fish and they will soon grow and reward you with many years of trouble-free fishkeeping.

When to buy your fish

Once you have dug out the pond and filled it with water, does this mean you are ready to add the fish? It may be tempting to go straight out and stock your pond, but there are a couple of matters to consider first.

Before adding any livestock, be sure to dechlorinate the water used to fill the pond. If the pond is made of concrete or if rendering or concrete have been used nearby, test the pH level to ensure that there are no lime residues. If there is any evidence that lime or masonry has fallen into the pond, you must drain and refill it. Then test it again.

Although you can add fish as soon as the water has been treated and tested, it is better to add some plants first. This gives them a chance to establish for a few weeks before the fish are introduced. In addition, the water will be able to reach a more balanced state during these two to three weeks, and it gives you a chance to adjust the pump and filter flow rates, as well as establishing that the pond does not leak!

The best time of year to add fish to the pond is between mid-spring and early summer. By mid-spring, fish have regained any weight they lost during the winter and water temperatures will have stabilized. Plant growth is rapid and provides the fish with shelter from predators

Stocking levels for your pond

There are no hard rules regarding the stocking levels for ponds, as there are so many different parameters: size and type of fish, quality of filtration, experience, turnover of water through the filter, feeding levels, etc. However, as an approximate guide, do not stock more than 60cm (24in) of fish body length (excluding the tail), per square metre of surface area. Do not buy the maximum stocking level of fish for your pond in one go. Instead, make several purchases over a period of time to allow the pond and filter to adjust to the increased biological activity caused by adding fish to the pond. Test the ammonia and nitrite levels regularly after introducing livestock to the pond to ensure that the filtration system is coping with the increased levels of feeding and waste production.

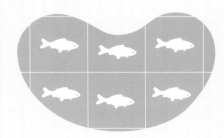

before they become acclimatized to their new surroundings. Do not buy your entire pond stock at once. Filters take time to establish, and introducing fish gradually over several weeks will allow the bacteria in the filter to multiply at a sufficient rate to maintain good water quality.

Before buying any fish, test your pond water for ammonia and nitrite (see page 16). If either test gives a result other than zero, wait until levels have dropped before adding any livestock to the pond. Buy your fish from a reputable dealer and only select healthy, robust fish. Choose well-rounded specimens with no skin damage. (The occasional small nick in the fins or a missing scale should not be a problem as long as there is no bacterial infection.) The eyes should be clear not cloudy, and fins should not be torn or bloodshot.

Fish to avoid
Never buy specimens from tanks containing other fish that frequently flick their sides as if trying to scratch an itch. This is a typical symptom of parasites and fish showing these signs should be avoided at all costs. Avoid any small, weak specimens that swim lethargically in the dealer's tank. Do not buy 'bargain' fish that look unwell or deformed. Such acquisitions can ultimately prove expensive, as they may infect other fish in the pond.

Where to buy your fish
If possible, buy fish locally from a specialist outlet. Such fish will need to endure less stressful travelling time and are likely to have been kept in similar water conditions to those in your own pond. When you have selected a fish, it will be transferred

Quarantining new fish to protect existing stocks

Although you choose your fish from apparently healthy stock, there is always the chance of a disease developing. This may result from the stress of transportation, or the fish may already be harbouring a low-level infection that manifests itself when they are caught and moved. It is therefore a good idea to quarantine any new purchases in an isolated environment for at least four weeks before adding them to a pond that already contains fish. Quarantine pond fish in an aquarium with a capacity of 100 litres or more. Furnish the tank sparsely and add a filter to maintain good water quality. Replacing about 25% of the water with pond water every week will allow the fish to become acclimatized to any changes in water conditions. Be sure to use a different set of equipment for the quarantine tank than is used for the pond to prevent the possibility of spreading disease. Providing the fish show no signs of disease after four weeks, you can safely introduce them into the pond.

to a plastic bag, usually containing one-third water to two-thirds air. If you have a long return journey, insist on having oxygen added to the bag instead of air. Oxygen will sustain the fish for several hours on the journey home.

When your fish is safely home, float the unopened bag on the surface of the pond for at least half an hour to allow the water temperature to equalize. Then open the bag and roll down the sides so that it floats on the water. Over the next half an hour, add water from your pond to the bag. Doing this reduces the stress caused by any differences in pH levels and water hardness. After this time, it is safe to release the fish from the bag.

Catching and handling fish
If you should ever need to catch your fish, for example in order to inspect them, treat them for damage or to move them to another pond, be sure

Putting hardy goldfish into a new pond

1 Take your fish home as soon as you can. The dealer will put them in a plastic bag with air (or oxygen for a long journey). Keep the bag cool and covered, as fish are less stressed in the dark.

2 As soon as you get home, float the bag on the pond with the top tied. After 30 minutes, open the bag and transfer some of the pond water into the bag. Do this over a period of 30 minutes.

3 Slowly tip the fish and water out into the pond, making sure that all the fish are out of the bag. The fish will hide for a day or two, so there is no need to feed them. After this, feed them once a day to start with, then twice a day.

to observe the following rules. Try not to handle the fish. Instead, use a soft hand net, large enough to hold the fish. If you need to use your hands to hold a fish still, wet your hands with pond water to protect the fish. Ideally, though, use a damp cloth. Make sure that it has not been treated with detergents, etc., as these can burn the fish's delicate skin. Keep the fish supported at all times. It is a good idea to have a bag ready, part-filled with pond water. If the fish cannot be coaxed into the bag, catch it carefully and lower it gently into the bag. As you remove the net, make sure the fish does not become caught in the mesh. If it does, take time to remove the delicate fins carefully from the mesh. Fish are often stressed when being caught and this, together with subsequent damage, can lead to infections.

When bagging fish for transportation, try to trap as much air in the bag as possible. Fish survive longer when the bag is filled with at least two-thirds air to one-third water. Do not blow into the bag to inflate it, as the resultant carbon dioxide is toxic.

A selection of pond fishes
On the following pages we take a look at the most popular types of pond fish available and the ponds that suit them.

Goldfish (Carrasius auratus)
Goldfish remain the most popular type of pond fish and are often the first ones to be added to a new pond. Their bright colours are only one

reason for their popularity. They are extremely hardy and will survive a wide range of temperatures and types of water. Add to this their low cost, long life span (20 years is not uncommon) and ease of breeding, and the goldfish is quite justifiably the ideal pond fish.

There are many strains of goldfish available today. The most common is the gold or red variety, often with black or white patches. Common goldfish may reach over 23cm (9in)

Below: A huge range of colours ensures that no two shubunkins ever look the same. These popular fish are a variety of the common goldfish and, being almost as hardy, are suitable for most ponds.

in length and will breed from 10cm (4in) upwards. Long-finned varieties of the common goldfish are also available; it is important to select fish with no fin damage.

For a little variety, shubunkins are an excellent choice. They are a multicoloured fish, with colours ranging from blue and white to black, red and yellow. Shubunkins are often speckled and have a variety of scale formations.

The fast, streamlined, long-tailed goldfish, or comet, is generally red. Its tail is extremely long, giving the fish its name. The comet is always active and prefers a reasonably large swimming area. Look out for the elegant red-and-white sarasa comet.

Fancy goldfish

There is a huge range of fancy goldfish varieties available. Some are suitable for ponds, but many more are best kept in indoor aquariums. Fantails are the most suited to ponds, but you may need to bring small fantails indoors when the temperature drops. These squat, short-bodied fish are slow swimmers that prefer being kept with fish of a similar size and speed.

More exotic than fantails are the varieties of fancy goldfish that originate from the Far East. Fish such as black moors with their protruding eyes, and pearlscales with their knobbly, golf-ball-like appearance, are not ideally suited to

Left: *Fancy goldfish, such as this fantail, are slow-swimming and prefer slow-moving water and smaller, well-planted ponds, away from more active fish and predators.*

Above: The most
popular pondfish –
the common goldfish
(Carassius auratus) – *is
robust, colourful and
easy to breed, making
it ideal for almost any
pond. Goldfish are
sociable fish, so keep
them in small shoals.*

Left: Sarasa comets are
an attractive addition
to any pond. Their
elongated tails and
active nature indicates
a preference for open
swimming space.
When buying comets,
check the fins carefully
for frayed edges, which
are a sign of finrot.

77

ponds. They may be kept outdoors in summer, but you must keep them in an indoor aquarium throughout the winter in temperate climates. Their shape prevents them from active swimming and they may miss out at feeding time when faster, more aggressive fish are present.

Golden Orfe (Idus idus)

These fish are extremely active and prefer well-oxygenated ponds with a good flow of water. Golden orfe are relatively hardy but susceptible to lack of oxygen, especially in hot weather. Orfe are not recommended for small ponds, or for ponds without a fountain or waterfall to provide good gaseous exchange.

Below: Although they are called golden orfe, these fish range in colour from yellow through to deep orange. Orfe will grow quickly in well-filtered ponds and reach a size of 60cm (24in) or more. A nervous fish, best not kept with koi.

Orfe are sometimes seen with blue coloration, instead of their usual orange form. These long, streamlined fish are happiest in small shoals and can often be seen removing insects from the pond surface. They are excellent jumpers and have been known to leap from small ponds.

Tench (Tinca tinca)

Golden and green tench are commonly available pond fish. They are instantly identifiable by their extremely slimy skin, which is covered in a thick layer of protective mucus. They prefer quieter waters, with less flow than many other species. However, they are extremely adaptable to most pond conditions and grow quickly. Green tench can reach 45cm (18in) in length. These excellent scavengers generally feed from the bottom of the pond and are often not seen for weeks at a time, while they pick up insects and other food deep down in the water.

Above: Green is the tench's normal colour. Tench are extremely hardy fish and will thrive in most ponds. However, due to their colour and bottom-dwelling habits, they are rarely seen.

Below: The golden strain of tench is a very attractive pond fish. In common with their green relations, golden tench thrive on a variety of foods, including worms and maggots, as well as dry food.

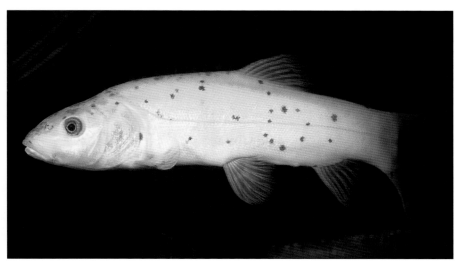

Golden rudd
(Scardineus erythrophthalmus)
Another popular coarse fish, the golden rudd, is silvery in colour and has a metallic appearance. Its fins are usually orange and the skin has a slightly rough appearance. Golden rudd may grow to more than 38cm (15in) long.

Rudd are best kept in small shoals and will adapt to most ponds. They scavenge at most levels in the water, as well as readily taking dried foods. In common with golden orfe, rudd prefer well-oxygenated water and frequently spawn in larger ponds.

Grass carp
(Ctenopharyngodon idella)
These often misunderstood fish are frequently considered ideal for keeping down levels of algae and blanketweed. However, while it is true that they do eat these unsightly plants, they do not restrict their diet to algae and soon devour waterlily stems and oxygenating plants. As they can reach a maximum size of 90cm (36in), grass carp are not suited to small, heavily planted garden ponds, but they are ideal for large koi ponds, which do not usually contain plants. In this environment, grass carp are an extremely hardy fish that perform a useful algae-removing function.

Sunfishes, Pumpkinseeds and Basses (Lepomis gibbosus)
These fish generally originate from North America and can provide a contrast in shape and coloration to more commonly found goldfish-type species. Some species are extremely predatory, so always seek advice before buying these fish.

Above: *Grass carp will not only eat algae, they will also break up and eat many pond plants. For this reason, they are best kept in larger ponds with little plant growth. They make good companions for koi in unplanted ponds.*

Left: *With their metallic scaling and orange-tinged fins, golden rudd provide a flash of colour in the pond. They thrive in shoals, in well-oxygenated waters. In good conditions, the fish soon multiply and many of the relatively large fry will survive. Keep these fish in a pond large enough to support their numbers.*

81

Koi (Cyprinus carpio)

Koi are the ornamental carp that many pondkeepers aspire to keeping. They are descendants of the common carp and have a well-catalogued history dating back hundreds of years. Although koi originated from Japan (the best-quality fish still come from there), they are now bred all over the world and fish are regularly exported from Israel, South Africa, the United States and most of Southeast Asia.

Koi have won the hearts of fishkeepers through the sheer variety of the colours available and their endearing characteristics. These fish can attain over 90cm (36in) in length and are easy to tame so that they can be fed from the hand. Ornamental koi share many of the strengths of their common carp ancestors. They are tolerant of a wide range of temperatures and water types and will eat almost anything!

Koi require a large pond to grow properly and will become stunted if kept in small or even average-sized garden ponds. Their capacity for food means that good filtration is essential. There are several excellent books available, specifically concerned with keeping koi, so consult these and seek the advice of a reputable dealer before considering keeping these beautiful fishes.

Right: Koi are always looking for food. Be sure to feed them with a good-quality diet, based around pellets or sticks. Spread the food around the pond to make sure that all the fish get some of it.

Left: There is a wide variety of gold and silver Ogon koi available. Koi are found in all colours, which is one of the reasons for their popularity. Keep them in well-filtered ponds with plenty of swimming space.

Sterlet (Acipenser ruthenus) and sturgeon

Sturgeon are known world-wide for their caviar, but the sturgeon offered for sale for garden ponds are not the giants of the Volga. They are sold at 5-7.5cm (2-3in) upwards and their prehistoric appearance makes them an appealing purchase. They can grow to around 90-120cm (36-48in). Clearly, therefore, they are not suited to any but the largest ponds.

Sturgeon and sterlets require a large quantity of specialized foods, as well as cold, well-oxygenated water. While they mix well with koi in specialist ponds, but they are not a good fish for the average pond.

Catfishes

There is a huge variety of catfishes around the world and they vary from small species less than 2.5cm (1in) long to voracious fish-eaters well over 120cm (48in) in length. Before buying catfish species, find out how large they grow and whether their natural diet is fish! In particular, avoid the blue or golden channel catfish *(Ictalurus)*, which are often offered for sale, but which grow extremely quickly and will find other fish in the pond an irresistible treat.

Below: Sterlets and sturgeon will grow quickly and require large, well-oxygenated ponds in which to thrive.

Below: Many catfishes are large, voracious predators and are not suitable for the garden pond. This is Ictalurus punctatus – *do not add it to your pond!*

Check before you buy

Recent increases in the variety of fish species available to pond-keepers have led to heightened threats to indigenous freshwater fish. Only buy legally available species and never release fish into the local water supply. Failure to observe this responsibility could lead to the decimation of natural wildlife, as new introductions may bring disease and predatory threats that local environments are not used to and unable to withstand.

UNDERSTANDING YOUR FISH

Understanding the fish in your pond, what makes them thrive, what distresses them, and how they grow and breed not only adds to the enjoyment of keeping them, but also increases the chances maintaining healthy stock.

Silvery protection

Scales protect the fish's soft tissue below and are themselves protected by a layer of mucus. Handle fish with care in order not to damage them.

Fish have evolved over many years to suit their particular environments. For example, a round, fat-bodied fish will not be at home in fast-moving streams. Many fish have down-turned mouths that help them as they search for food at the bottom of rivers and ponds. Fish that inhabit muddy environments often have barbels around the mouth in order to detect their food.

The swimbladder
Most fish have a buoyancy device called a swimbladder. This is an adjustable airbag that allows fish to remain at various depths in the pond.

The gills
The gills are located just behind the head and are protected by a gill cover plate called the operculum. The gills allow the fish to expel waste products, such as ammonia

and carbon dioxide. This process is similar to breathing and works by passing water from the mouth through the gill filaments and out through the operculum. As water passes through the gills, it comes into contact with the tiny blood vessels that absorb the oxygen from the water and then expel the toxins. Damage to the gills is extremely serious, although fairly easy to detect (see page 101).

Fins

Fins are used both for stability and propulsion. In addition, many fish species use them during courtship and to fan their eggs. Most fishes have seven fins, although some species have evolved different fin structures to suit their own habitats.

The caudal fin (or tail) is mainly used for propulsion, while the dorsal and anal fins act as stabilizers, almost like the keel on a boat. Two pairs of

fins – the pelvic and pectoral – are used for directional control and adjusting the position of the fish in the water. The fish uses these fins for turning and stopping. In fish that inhabit fast-flowing waters, these fins develop far more strongly than in species found in slow-water habitats. It is important to check the finnage of fish in your pond regularly, as the soft tissue can be easily damaged and infected during squabbling or courtship behaviour.

Scales

The scales are covered in a protective layer of mucus that helps keep the fish free from infection. The scales themselves grow constantly and protect the soft tissue beneath.

Below: This view from above is how we normally enjoy the fish in our ponds. These fish frozen by the camera show various aspects of their form and function.

The stabilizing dorsal fin.

The pectoral fins are swivelling independently to alter the fish's position in the water.

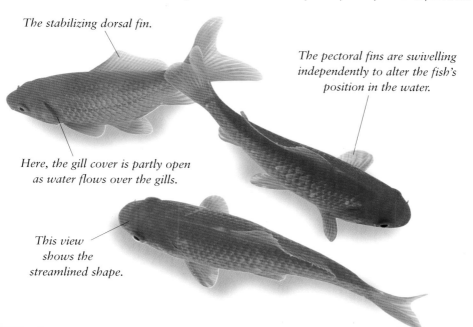

Here, the gill cover is partly open as water flows over the gills.

This view shows the streamlined shape.

The anatomy of a typical fish

Lateral line
The line of dots seen along the flanks are tiny holes leading to a canal that runs the whole length of the fish. Nerve endings in this respond to changes of water pressure and enable the fish to detect vibrations in the water.

Dorsal fin
This single fin acts as the main 'keel' to prevent the fish rolling during swimming.

Caudal fin (tail)
Most fish swim by sweeping the back part of the body from side to side. The tail fin helps to convert that movement into forward thrust.

Anal fin
This single fin acts as a stabilizer during swimming and in males of certain fishes is modified for breeding purposes.

Vent
This is the opening of the digestive, urinary and reproductive systems.

Scales
These are extremely thin, overlapping bony plates that protect and streamline the body. They contain pigment cells, although apparent colour in many fishes is produced by reflected patterns of light.

Eye
Sight is well developed in most fish and the eyes are most sensitive in yellow-green – those wavelengths that penetrate furthest into water.

Nostril
Many fish have a highly developed sense of smell. The nostrils do not connect to the mouth.

Mouth
The shape and position of the mouth varies between surface, midwater and bottom feeders. This is the upturned mouth of a surface feeder.

Pelvic fins (paired)
These also help the fish to control its position.

Pectoral fins (paired)
Fish use these to help steer and change their position in the water – even to swim backwards!

Operculum (gill cover)
Water taken in through the mouth is forced out over the blood-rich gills beneath this cover. In the process, oxygen is absorbed and waste products released.

The senses

Fish rely on the same five senses that humans possess. The main difference is that fish use their nostrils only for smelling or detecting food, not for breathing. In addition to being able to hear, fish have an extra sensory faculty that detects underwater vibrations. This is known as the lateral line and is visible on most species as a small row of holes along the body of the fish.

The digestive system

Most pond fish are omnivorous (they feed on both plant and animal materials). Food is taken in through the mouth and is broken up by the pharyngeal teeth. The process of digestion begins as food passes through the fish's digestive juices. The long intestines help to digest fibrous plant material and enable the fish to dispose of indigestible material through the vent.

Left: Koi will eat a wide variety of food. Use a good-quality koi stick food as a staple diet in spring and summer. Koi are naturally inquisitive and quickly become hand-tame at feeding time. This provides an ideal opportunity to check them over for fin and scale damage.

Once the fish have grown to 5cm (2in) or more, offer them sticks or pellets that provide more food per mouthful than flake.

Pond flake is ideal for small fish up to about 5cm (2in), as they will find it easy to digest.

Specialist foods improve colour and encourage growth. Feed rates vary, so follow the directions.

*Right: Frozen foods
are convenient to use
and ideal as a treat
for small fish or for
bringing fish into
breeding condition.*

*Thoroughly defrost the
bloodworm before adding
it to the pond.*

*Frozen bloodworm is
packaged in small
blocks.*

Feeding your fish

There is a wide variety of dried foods available for pond fish. They are commonly found in stick, pellet and flake forms, with the stick variety being the type of food normally preferred for an 'average' pond. Food sticks tend to float initially and then sink as they soak up water. Good-quality sticks will not break up too quickly and will therefore allow fish feeding at all levels in the pond to obtain a regular source of food.

If using dried foods, seek the advice of your specialist aquatic centre to make the right selection. Find out which foods they use, and why. Listen to their specific recommendations, as the cheapest food is rarely the best. A good, balanced food will contain all the ingredients required for fish to grow and remain healthy. The very best foods use fish protein as a part of the diet. This aids digestibility and naturally enhances both growth and colour in your pond fish.

If your pond is stocked with small fish, consider mixing some flake food in with the sticks, to ensure that the smaller fish can feed. If your pond contains specialist bottom-feeders, such as tench, sterlets or catfishes, provide them with a balanced diet by adding some sinking pellets to the regular stick or flake food mix. In addition to dried foods, many shops also sell live, natural foods such as daphnia and bloodworm. Provided these are from a clean source, they are worth feeding on occasion as a treat. These foods are also available in frozen and freeze-dried forms. Within time, it is likely that your pond will cultivate its own supply of live foods.

It is advisable to feed your fish during the day, rather than in the late evening. This is because fish are ectothermic (cold-blooded) creatures and their metabolism slows as the temperature drops. Feeding them during the day will ensure the digestive system is working at its most efficient.

Feed the fish sparingly two or three times each day throughout the spring and summer. Put only as much

91

food in as the fish will eat within five minutes and remove any uneaten food. Do not be fooled by the always hungry appearance of pond fish. While they will continue to eat almost as much as you are prepared to feed them, their digestive systems cannot cope with large amounts of food. Any food in excess of their needs for growth will be passed quickly through the digestive system and may pollute the pond.

Remember to reduce feeding levels as the water temperature drops through the year. Fish will feed until the water temperature drops to around 5°C (40°F), but they will consume less food. They do not fully hibernate through the winter and because the digestive system of most pond fish continues to function, it is a good idea to feed them a low-temperature food in the winter. These foods are more easily digested in colder conditions. Most low-temperature foods contain wheatgerm to aid digestibility, and the very best low-temperature foods also contain a high level of fish protein as this is far more easily converted into body mass.

In winter, feed the fish once daily and ensure the food is being eaten. If not, cease feeding for a couple of days and remove any uneaten food from the pond.

The kidneys

Freshwater fish do not drink water. On the contrary, they must constantly expel excess water and with it, all the unwanted waste products, salts and urine. This essential function is performed by the kidneys. They are relatively large in most freshwater fish, as water needs to be expelled at the same rate at which it is drawn in through the body tissue.

Life expectancy

There is great variation in the life expectancy of fish species. In the correct conditions, koi and sturgeon may live for decades and can attain a considerable size. Koi are often seen at over 60cm (24in) long, and can attain nearly 1m (39in) if given room to grow. Goldfish can be expected to live for up to 20 years and can reach up to 25cm (10in) in length, whereas the more fancy varieties have a life expectancy of about four to five years and may be fully grown at 10-13cm (4-5in). With any species of fish, the key to a long and healthy life is to live in the correct, disease-free environment with the right feeding regime.

Fish breeding

Most pond fish spawn during the late spring and summer months, when conditions are ideal for breeding and raising fry. The water is warm and plants provide natural protection for young fish. There is an abundance of natural food and adult fish are in the best of health, having fed all spring and thus increased their body weight, vitality and colour.

Spawning in spring or summer means that newly hatched fry are able to grow and feed before the onset of winter. This enables them to build up the necessary reserves for

the months ahead, when the water temperature drops and food becomes less readily available.

The first signs of spawning activity
Spawning is a time of frenzied activity in the pond. It may appear that some fish are being harassed by others, as a mad chase ensues around the pond. However, it is more likely to be spawning activity than aggressive behaviour and will culminate in thousands of tiny eggs being deposited on pond plants.

Differentiating the sexes
During the breeding season, it is quite obvious to the eye which of the sexually mature fish are females and which are males.

The female fish in the pond become plump and rounded and differ markedly from the males, who remain sleek and streamlined.

Above: The white pinhead-sized tubercles are clearly visible on the gill plate of this male shubunkin. These 'pimples' play a role in helping male fish stimulate females into a mating response.

While females become more rounded, mature males develop small pimples on the gill covers and pectoral fins. These 'tubercles' slightly resemble the whitespot parasite, but should not be mistaken for it. If handled, the tubercles feel rough; the male uses them to rub against the female during breeding. It is far easier to sex pond fish correctly during the breeding season, as males will be chasing females around the pond with great enthusiasm.

Interbreeding
Most goldfish colour variants will interbreed, as they are all one species. This can result in weak offspring

being born, but is unavoidable in the pond. In most cases, the pondkeeper does not realize that spawning is taking place until it is well advanced!

Controlled breeding

The only way to control spawning and ensure the best possible fry is to select the parent fish and spawn them in an aquarium. When selecting prospective parents, choose two or three males for each female. Examine the fish from above, as the difference in shape is most easy to see from this angle. The parent fish should have no deformities and be in the best of health, as spawning can be stressful.

Feeding before breeding

Correct nutrition makes a major contribution to the vitality and colour of a fish, so supplement the normal diet of fish in a breeding aquarium with good-quality frozen foods, such as bloodworm and daphnia. Feeding these prepared foods will make up for any live foods that the fish are not eating while they are in the aquarium.

The breeding tank

Of course it is important to provide the parent fish with good water conditions in which to spawn. Use an aquarium with a capacity of 50-100 litres and equip it with a good-quality filter that has been matured for some time. Fill the tank initially with an equal mix of pond water and conditioned tap water. Carry out water changes as required to maintain the water quality, using dechlorinated tap water. Ideally, leave the aquarium bare, except for bunches of pond plants, such as myriophyllum, elodea or other bushy

plants. Alternatively, use spawning mops or mats instead of plants.

Spawning activity

When spawning takes place, the male shepherds the female into the plants or spawning mops. The female will deposit the eggs amongst the plants, where they will be protected against other adult fish. After spawning, remove the adult fish, as they may eat the eggs. Alternatively, transfer the eggs, still on the plants, to another aquarium. Be sure to keep the eggs submerged during the transfer. If possible, keep the adult fish away from the pond for a few days, so that they can regain their strength and vigour.

Hatching and care of the fry

Goldfish eggs hatch after about 96 hours. When the fry are first born they do not require feeding, as they will use their yolk sacs as a source of food. Offer the first feed when the young fry are free-swimming. Using a proprietary fry food is a lot less time-consuming than cultivating infusoria and similar small foods.

Be sure not to overfeed the fry; this can lead to internal problems, as well as causing the water quality to deteriorate. It is important to keep ammonia and nitrite levels at zero and to allow the small fish as much space as possible. This may involve setting up additional aquariums, although regular culling of any weak or deformed fry should ensure that only a few, strong fish survive.

Leave the fry in the aquarium until they are at least 2cm (0.8in) long to increase their chances of survival once you place them back into the garden pond.

Left: Spawning is a vigorous activity that can lead to skin and fin damage. These are koi spawning amongst artificial mops placed in the pond.

Right: This close up shows the eggs caught in the strands of the spawning mop, thus protecting them from the rigours of pond life and increasing the likelihood of a successful spawning.

FISH HEALTH

Here we look at the most common fish diseases and how to diagnose and treat them. Correctly identifying the problem is the key to successful treatment. There are many excellent remedies, but they are ineffective if the diagnosis is wrong.

A regular inspection of your fish at feeding time will often alert you to the first signs of a problem. Fish may look listless or have difficulty breathing. There may be physical signs of ill-health or a fish may not appear at feeding time. Recognizing these early signs as potential problems means you can swiftly implement the appropriate treatment.

Testing the pond water
As we have seen, good water quality plays a major role in maintaining the health of your fish, so test the pond

A good start

Buying healthy specimens from a reputable dealer is the first step to a flourishing pond. Keeping it that way calls for vigilance and prompt action.

water regularly, particularly for ammonia, nitrite and pH levels. If fish are showing signs of sickness or distress, testing the pond water should be your first course of action.

Listless fish or fish gasping at the surface may be suffering from a lack of oxygen or a high nitrite level, both of which can be rectified without adding chemicals to the water. Alternatively, these symptoms may be caused by gill infections or other diseases. The important thing is to eliminate water quality as a problem before proceeding further. If there is no problem with the water, you can begin to consider other options. Here, we look at the most common fish diseases and how to diagnose and treat them.

Skin parasites

Parasites are usually easy to detect, as they are often visible to the naked eye. Even before you see the pinhead-sized parasites, you may notice fish flicking and rubbing against stones and plants in an effort to dislodge the parasites. This should be the cue for a closer visual inspection.
Treatment: Use a proprietary anti-parasite treatment.

Whitespot is the most common skin parasite and easy to treat if recognized early. Diagnosis is simple:

Above: You can buy anti-parasite pond treatments to tackle a range of problems. Containers with compartments holding a measured dose help accurate application.

the skin and fins of the fish will be covered in small white cysts, almost like a dusting of grains of salt. The cysts measure up to 1mm (0.04in) in diameter and spread rapidly across the fish if not treated. Fish may flick against rough surfaces to try to scratch off the parasites and this can often lead to secondary bacterial

Hygiene and safety

For your own safety, do not handle fish if you have open cuts or abrasions. Whenever testing water or performing maintenance around the pond, ensure that you always thoroughly wash your hands afterwards. Diseases such as fish TB can be passed to humans and will result in severe discomfort. To prevent diseases spreading from a

quarantine tank to the pond, avoid using the same equipment for both. Keep nets separate and be sure to wash your hands after handling sick fish. If on occasion you need to use the same equipment for both the treatment tank and the pond, clean the items using one of the many proprietary disinfectants available from your local aquatic specialist.

infections. In the case of heavy infestations, gill movement will increase as breathing becomes more difficult. Infected fish often lie listlessly in shallow water.

Whitespot is highly contagious and multiplies more rapidly in warm conditions. The cysts, which feed on the tissue of the fish, fall to the bottom of the pond and divide hundreds of times. New parasites swim off to search for a host, but die if they cannot find one.

Cysts cannot be treated while attached to the fish, as they live beneath the skin, so chemical treatments are aimed at the free-swimming stage of the life cycle and work quickly and effectively.

Treatment: Ask your specialist aquatic shop to recommend a proprietary whitespot treatment and be sure to administer the full course. Although cysts quickly disappear from the host fish, the treatment must be allowed to affect the free-swimming parasites.

Below: Whitespot cysts are clearly visible as pinhead-sized spots on this tench. Whitespot can infect most species of pond fish, but responds to prompt treatment.

White spot

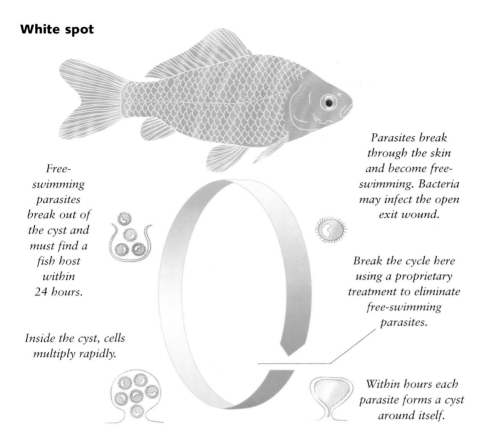

Free-swimming parasites break out of the cyst and must find a fish host within 24 hours.

Inside the cyst, cells multiply rapidly.

Parasites break through the skin and become free-swimming. Bacteria may infect the open exit wound.

Break the cycle here using a proprietary treatment to eliminate free-swimming parasites.

Within hours each parasite forms a cyst around itself.

Alternatively, leave the pond without any fish in it for 7-14 days. This is usually enough to ensure that the cysts die off, as there are no hosts to which they can attach themselves.

Slime disease is caused by a multitude of parasite infections and usually manifests itself by causing excess, greyish mucus on the body of the fish. They may rub against rocks and other rough surfaces in order to rid themselves of the irritation. *Treatment: Treat the pond with a proprietary anti-parasite treatment.*

Fish lice (*Argulus* sp.) are most commonly found on recently imported fish. It is important, therefore, that you ensure any fish you buy have been well rested and inspected by the dealer.

Fish lice are easily visible to the naked eye. They are 5-10mm (0.2-0.4in)-long, greenish, disc-shaped parasites. Fish lice grip on to a fish with two strong suckers and draw blood through the victim's skin. They can be free-swimming and will cause distress if left untreated. *Treatment: First remove all visible*

Fish lice

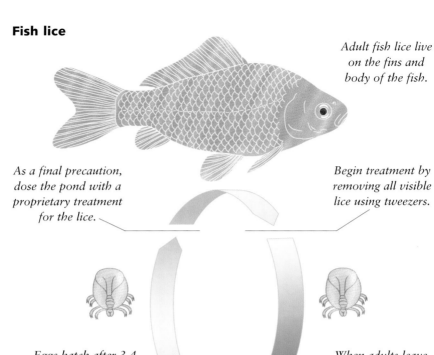

Adult fish lice live on the fins and body of the fish.

As a final precaution, dose the pond with a proprietary treatment for the lice.

Begin treatment by removing all visible lice using tweezers.

Eggs hatch after 3-4 weeks. The lice then begin to mature into adults, attaching themselves to fish.

When adults leave the host to lay eggs, they leave open wounds.

Eggs are laid in long strings, and stick easily to plants and rocks.

Right: *Adult fish lice (Argulus sp.) can be clearly seen when they attach themselves to the body of a fish, here a goldfish. Treatment involves detaching the adult lice, tending the resulting wound with an antibacterial product and dosing the pond to eliminate any remaining lice.*

lice. This is best done with a pair of tweezers. Treat the areas where the lice were attached with an anti-bacterial treatment and finally dose the pond with a proprietary treatment for lice and flukes.

Gill flukes cause respiratory distress in fish, which are often seen gasping at the surface of the pond. If you carefully open the gill covers, you will be able to see the tiny 1mm (0.04in)-long wormlike flukes. *Treatment: Gill flukes can be removed by treating the pond with a good-quality anti-parasite treatment, but you must catch the infestation*

early to avoid irreparable damage to the gills of affected fish.

Other gill problems can be caused by poor water quality during shipping. Fish either lie motionlessly at the bottom of the pond or will be seen gasping at the water surface. Inspect the gills and if no parasites are apparent, simply ensure that the water quality is good and that there is plenty of oxygen. The fish should quickly start to recover.

Anchor worm parasites are easily visible to the naked eye, growing to more than 2cm (0.8in) and attaching

Gill flukes

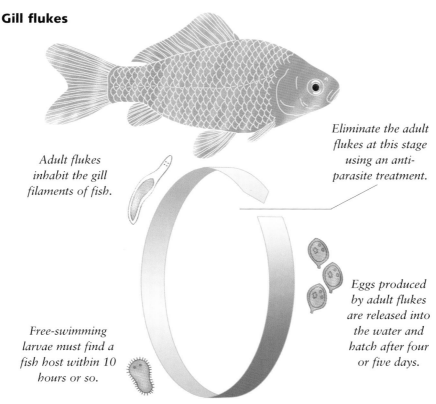

Adult flukes inhabit the gill filaments of fish.

Eliminate the adult flukes at this stage using an anti-parasite treatment.

Eggs produced by adult flukes are released into the water and hatch after four or five days.

Free-swimming larvae must find a fish host within 10 hours or so.

Anchor worm

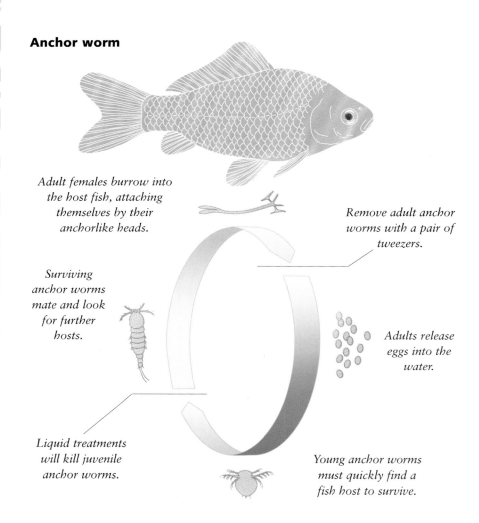

Adult females burrow into the host fish, attaching themselves by their anchorlike heads.

Remove adult anchor worms with a pair of tweezers.

Surviving anchor worms mate and look for further hosts.

Adults release eggs into the water.

Liquid treatments will kill juvenile anchor worms.

Young anchor worms must quickly find a fish host to survive.

themselves to fish by burrowing their heads into the body. The affected area becomes raised and reddened and susceptible to bacterial infection. *Treatment: Before treating the pond, remove as many anchor worms as possible. Using a pair of tweezers, grip the parasite as near to its point of entry as possible and gently pull it free from the fish. If you pull out the whole worm, the anchorlike 'head'* *will be visible. Treat the wound with an antibacterial treatment and dose the pond with a proprietary anti-parasite treatment.*

Finrot
Finrot is caused by harmful bacteria and easy to diagnose. The fins or tail take on a ragged appearance as they are attacked by the bacteria. In severe cases, only the fin rays may

Above: This is an extreme case of finrot on a bubble-eye goldfish. Finrot causes the degeneration of the membrane between the fin rays. Fatal if not treated.

Below: When removing adult anchor worms from the body of a fish, be sure to detach the 'head' as well. Work carefully and treat the wound afterwards.

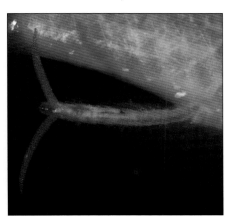

remain. The bacteria are always present in the water and only manifest themselves when fish become stressed, often due to less than ideal water conditions. Female fish may suffer from finrot during the summer when they are weakened by spawning activity.

Treatment: If fish show signs of finrot, first test the water to ensure that conditions are good. Follow this with a course of antibacterial treatment. Be sure to treat the whole pond, as well as isolating any fish showing symptoms of finrot.

Ulcers

Ulcers are caused by various bacteria and appear first as reddened areas on the body or at the base of the fins. These quickly develop into open sores that appear to eat into the fish, often exposing the skeleton, before

Above: Ulcers cause reddening at the base of the fins and vent and can spread quickly. Isolate affected fish and treat them with a proprietary anti-ulcer remedy and improve the pond conditions.

Right: A fish displaying signs of difficulty remaining balanced in the water. This is a typical symptom of a swimbladder infection. Treat the condition with a proprietary product and a tonic salt.

the fish die. The bacteria released by sick fish can cause ulceration to spread rapidly in heavily stocked ponds. Ulcers can also lead to cottonwool-like fungal growths around infected areas. Maintaining good water quality is one way of reducing the chance of ulcer disease, as infected fish are often stressed through poor water quality, poor handling during shipping or overstocking.
Treatment: Treatment is relatively simple, provided the disease is caught early. Try to isolate infected fish and improve the water quality. Use a proprietary anti-ulcer treatment. Be sure to treat both the isolated fish and the pond, even if other fish show no similar symptoms.

Swimbladder infections
Swimbladder infections may be the result of a variety of causes, the most common being either sudden changes in temperature or internal bacterial infections. Fish suffering from swimbladder disorders may appear outwardly healthy, but have difficulty maintaining their equilibrium in the pond. Affected fish often swim at peculiar angles and may even be seen upside down. Swimbladder problems are particularly common in fancy goldfish, as their unnatural body shape distends the swimbladder.
Treatment: Fortunately, there are now some proprietary swimbladder treatments available from aquatic shops. These generally work best if they are used in conjunction with a

tonic salt. The salt reduces the osmotic pressure on the fish, thus reducing the rate at which the kidneys have to perform, and greatly improves the chance of recovery.

Pop-eye

Pop-eye is rarely infectious. It affects usually one, or perhaps two, fish in the pond. As the name suggests, one or both eyes may protrude further than normal. Do not confuse this condition with the appearance of some of the fancy goldfish breeds, such as black moors, in which protruding eyes are normal. If the

symptoms spread to other fish, water quality may be at fault, so check the water at the first signs of pop-eye in any fish in your pond.

Treatment: *Treat affected fish with a broad-spectrum antibacterial remedy and provide the best possible water conditions.*

Fish tuberculosis

Fish TB may be harboured by many specimens without them showing any symptoms or distress. Problems often coincide with stress or poor water quality. When the infection does become apparent it usually acts very quickly. Fish suffering from TB quickly take on a hollow, emaciated appearance. Secondary symptoms, such as pop-eye and body damage, may occur.

Treatment: Fish TB is fatal unless treated and the few treatments available are expensive and often involve antibiotic injections. It is therefore important to ensure that when choosing stock for your pond, you select the healthiest-looking, most active and vibrant specimens. Do not buy runt fish or 'sad-looking' fish out of pity. If there are emaciated fish in the pond or tank you are selecting from, move on to another tank.

Immediately isolate fish showing symptoms of TB from your pond. Provide the best possible water quality and monitor the health of the affected fish. If it shows no sign of recovery, it is best to destroy it humanely.

It is now known that fish tuberculosis can be transmitted to humans. Be very careful, therefore, not to handle fish or equipment if you have cuts or broken skin. If fish TB is suspected, disinfect all equipment with diluted bleach, then rinse it until clean and leave to dry.

Dropsy

Dropsy is easy to identify, as affected fish take on a bloated appearance, with protruding scales like a pinecone. The eyes may also bulge outwards and the vent may become reddened. Dropsy occurs most often when water quality deteriorates. It may be that just one fish shows

Above: The protruding eyes of this koi may be caused by fish tuberculosis. Similar pop-eye symptoms can be the result of water quality problems.

symptoms or the whole pond stock may be affected. There are several possible causes of dropsy. The most common are bacterial infections and incorrect nutrition, although some viral infections give rise to the same disease.

Treatment: Because diagnosing the cause of dropsy is difficult, treatment is not easy either. It is best to isolate the affected fish quickly and provide them with the best possible water quality. Foods laced with antibiotics often prove successful, but they are expensive. In larger fish, it is possible to draw off the excess fluid that causes the scales to protrude, but this

Below: *Protruding scales and listlessness are signs of dropsy. Isolate fish as soon as you detect any symptoms. Monitoring your fish on a regular basis should alert you to potential problems in the pond.*

Above: Fungus has occurred on the tail of this fish, following an attack of finrot. If you treat skin damage promptly, fungal infections can be avoided.

should be done only by experienced professionals. If the condition worsens, the fish may have to be humanely destroyed.

Fungus

Fungus usually occurs as a secondary infection, following skin damage caused by finrot or flukes. Cottonwool-like growths appear on the affected areas. These growths may attract algae, giving them a greenish appearance.

Treatment: Treatment is usually effective if the fungus is caught early. Treat the pond with a proprietary anti-fungus treatment and monitor the fish for further outbreaks.

Carp pox

As the name suggests, carp pox is a viral disease that mainly affects carp, and overcrowding appears to be a contributory cause. Carp pox is readily identifiable, as waxy growths appear on the body and fins. These growths can reach 3cm (1.2in) in diameter but are rarely fatal. *Treatment: No treatments are currently available for carp pox, although disinfecting pond*

Disposing of sick fish humanely

On occasion, it may be necessary to destroy terminally sick fish. If you find the prospect distressing, consult a veterinary surgeon, who will ensure that the fish has a painless death. Should you have to destroy a fish yourself, the quickest and most humane method is to decapitate it using a sharp knife (or a pair of scissors with very small fish). Larger fish should be stunned first, with a blow to the head.

Ideally, dispose of the fish by wrapping it in several layers of newspaper and then in two or three plastic bags, before placing it in a secure bin. Alternatively, wrap it in newspaper and incinerate it.

equipment and filtration at the end of the season usually prevents the disease from returning.

The conditions discussed here are by no means the only ones found in pond fish. However, they are the easiest to identify and the symptoms described should help you to diagnose many common problems.

Before treating the pond or fish with any chemicals, be sure of your diagnosis. If necessary, consult your local aquatic specialist and always follow their recommendations.

Below: Carp pox is unsightly, but rarely causes great discomfort to the host fish. It may disappear as promptly as it appears.

CLEANING OUT THE POND

As with any part of the garden, pond plants will need to be trimmed and thinned to remove excess growth. From time to time it will be necessary to empty out the water to remove the build-up of sludge on the pond base.

Time for some remedial work

When plants have taken over the pond to such an extent that the water is barely visible, you must take urgent action to restore the situation.

'How often should I clean out my pond?' This is the question most often asked when it comes to pond maintenance. The answer is that it varies from one pond to another; smaller ponds generally require more frequent cleaning, as the plants outgrow the space much more quickly. An easy way to check whether the pond needs cleaning out is to run a net over the bottom of the pond. If it is full of half-rotted leaves and mud when you lift it out, you know it is time to clean the pond. The main reason that ponds require

cleaning is due to the liner. In a natural pond, any plant material would fall to the bottom and rot away into the soil, where plants could access the nutrients. However, in a pond with a manmade liner, there is less bacterial activity to break down the debris, so the process is much slower and the build-up of sludge on the base is quicker. The more debris there is in the pond, the lower the oxygen content will be. During periods of warmer weather, this will have a damaging effect on the fish and in certain conditions can kill all the fish. Clearly, therefore, it is important to keep the waste level low to avoid this risk. A biofilter will help to extend the period between cleaning but is not a replacement for all maintenance.

The other reason for emptying the pond is to remove excess fish, as most ponds tend to become overstocked. An opportunity to reduce numbers is always welcome, especially for the fish.

The second most common question is 'When should I clean out the pond?' The best time to do this is in early spring, before the frogs and other animals lay their eggs, or in late autumn after the first frosts. Frosts will kill off any foliage and makes cutting back easier. Avoid the summer period, as this is more likely to upset the fish and the balance of the pond.

As well as preparing any necessary tools and materials, you will need to have a holding tank ready for the fish you remove from the pond during the cleanout. It should be big enough to support the numbers of fish taken from the pond. A large, smooth, plastic container, such as a child's paddling pool, is ideal. As you empty the pond, do not pump muddy water down the household drain as it may become blocked.

Below: Cleaning out a pond can be a major undertaking, even if it is small. Allow enough time; until you start, you do not know how much there is to do.

Clearing out an overgrown pond

Before you start a major clearout, assemble some new planting baskets, a sharp knife, a supply of aquatic potting mixture, gravel and a liner repair kit, just in case of accidents.

If a particular plant tends to take over your pond, it may be worth replacing it with a dwarf variety or a different plant entirely.

Check for loose rocks or paving around the pond and secure them before starting on any work.

Before emptying out the pond or doing any work in the water, be sure to disconnect all electrical items, such as pumps, ultraviolet sterilizers, etc.

1 Remove any pumps and pipework and put them in a safe place away from the pond. Keep pets and children away from the site, as they may fall in when the pond is empty.

2 Set up a holding tank for the fish. Place it in the shade, so that the sun does not heat the water and upset the fish. If you intend to leave the fish in the holding tank for a few days, you will need to install a small pump to oxygenate the water. Put some pond water in a bucket ready to transfer fish from the pond to the holding tank.

3 Remove all planting baskets before emptying the pond. This allows fish to escape into deeper water when you reduce the water level. Check that no fish or frogs are caught up in any plant roots.

4 Lower the pump, but not by the cable. Drain the pond slowly, ensuring that no pond life is sucked into the pump. Transfer some water to the holding tank and the rest onto a nearby flowerbed.

Filling the holding tank

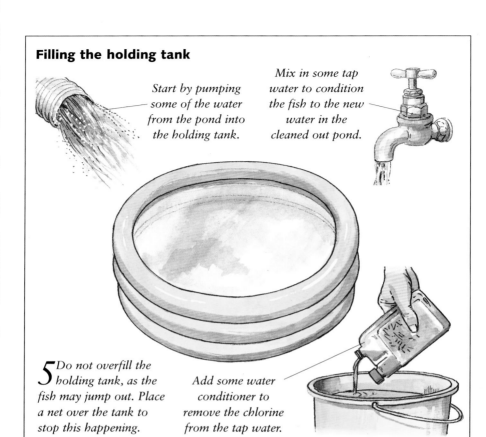

Start by pumping some of the water from the pond into the holding tank.

Mix in some tap water to condition the fish to the new water in the cleaned out pond.

5 *Do not overfill the holding tank, as the fish may jump out. Place a net over the tank to stop this happening.*

Add some water conditioner to remove the chlorine from the tap water.

6 *As the water level drops, stand the pump in a pond basket. This will prevent the pump sucking up animals and becoming blocked with leaves and debris.*

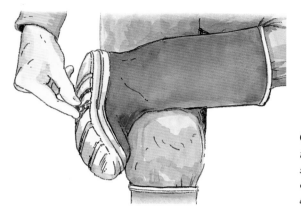

7 Before you step into the pond, check your boots or shoes for stones that could puncture the liner. Disconnect the pump before entering the water to avoid the risk of an accident.

8 Pick up the fish with your hands, as a net will scrape off the protective mucus on the fishes' skin. Place the fish into the bucket of pond water. This will wash away any mud on the fish.

9 When you have collected all the fish, remove them from the bucket and place them in the holding tank. Put the bucket right into the water to ensure that the fish do not fall into the pond from any height, as this could damage them.

10 Remove any stones or water-lily pots from the base of the pond. Using a dustpan (make sure it has no sharp edges that could damage the liner) and brush, sweep up the sludge from the pond and dispose of it. (It makes a very good fertilizer for the garden.)

Use your finger or a hose jet to direct the water and wash down the liner.

11 Hose down the pond with fresh water and empty it again. Wash as much of the mud as possible to the centre and then pump out the pond again. Check the folds in the liner for stones and remove any that you find.

Check dosages before using any water treatments. Follow any safety instructions on the container.

12 Add dechlorinator to a bucket of tap water and mix it well to achieve the right concentration for the finished pond volume according to the maker's directions. Pour the diluted dechlorinator into the pond as it refills.

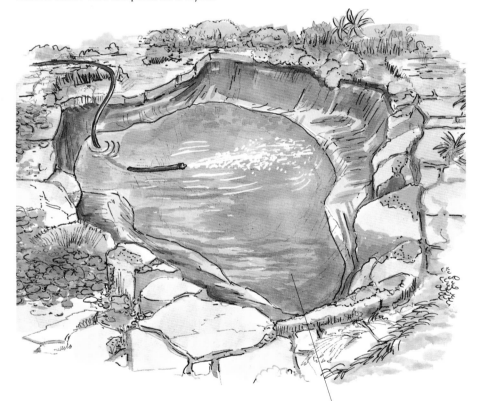

13 Place the hose in the pond and add water so that you can return the fish to the pond as soon as possible. When filling the pond, use a spray to help disperse the chlorine and warm the water.

Creating the best possible water conditions will mean less stress for the fish when you transfer them back into the pond.

117

14 To help you remove plants from their baskets, cut the roots back to the basket using a sharp knife. Do this away from the pond to avoid dropping the knife and damaging the liner.

15 Cut back iris. Reduce the height of the foliage by two-thirds and the length of the roots by half. Cut back the old flower stem to the main rhizome. These plants benefit from hard pruning.

16 Replant in good-quality aquatic soil, topped with a 2-3cm (0.8-1.2in) layer of rounded gravel. You can underplant tall plants with surface-covering varieties, such as creeping Jenny or cotula, as they will not compete with each other.

17 Water the baskets before placing them in the pond to wash away excess soil. To get the best from your plants, fertilize them after they have flowered, as the excess nutrients often stop flowering and encourage foliage growth instead. Place the planted baskets back into the pond, making sure that there are no stones lodged underneath them.

Repotting water plants

When repotting plants, remove any excess roots and cut back the remaining ones to about 15cm (6in). When you replant irises, leave rhizomes resting on the soil surface, but bury other plants up to the base of the shoot. In the case of waterlilies, cut back any black roots and leave the white ones. Cut back any soft, dead areas of rootstock to fresh new growth.

As you replace the fish in the pond, take this opportunity to check them for disease and wounds. Remove any excess fish.

18 Move the fish back to the pond once all the planted baskets have been replaced. (This avoids the risk of squashing the fish!) Scoop the fish out of the tank and place them into the pond.

The refurbished pond

Do not be tempted to overplant the newly restored pond. Keep any surplus baskets in a greenhouse in case some of the repotted plants do not survive.

Planting different species in separate baskets makes future pruning and control easier.

Leave gaps behind the baskets for the fish to swim through. If the gap is not sufficient they may get stuck and perish.

Always put plants that will grow tall in the largest pots to stop them blowing over when their foliage catches the wind.

In a newly replanted pond, you may have to provide some additional cover for the fish in winter. Old chimney pots or a large-diameter plastic pipe make good hiding places.

Pumps and filters

If your pond includes a pump and filter setup, you should clean the pump monthly and the filter completely in the spring and autumn. This is in addition to your regular maintenance around the pond. To clean a typical box filter, remove all the media, keeping a small amount to one side in some pond water to restart the bacteria. Hose all the filter mats/foam sheets and the plastic box with fresh water. Rinse well and replace. In the winter you may decide to turn off the pump and filter. In this case you should also make sure that the filter is empty of water. Take the pump apart following the manufacturer's instructions, clean it in water and reassemble it. Never use any detergents. Replace any seals or 'O' rings that are cracked or damaged. You can buy replacement parts from your aquatic dealer. Pondkeepers often overlook the pipework connecting pumps and filters, but this needs cleaning, too. Thread a length of string through one end of a piece of sponge, pull it through to clean the inside of the pipe and repeat as necessary. Rinse the pipe with fresh water, otherwise the waste will wash into the pond.

Cleaning a pond pump

Pond pumps are designed so that you can take them apart for cleaning and maintenance.

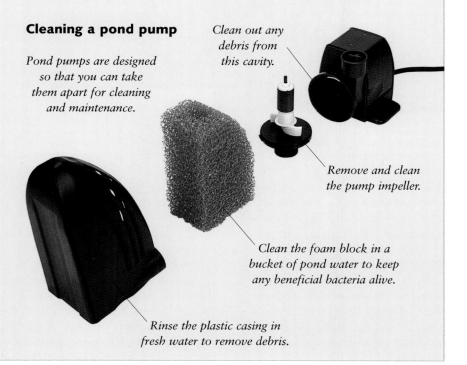

Clean out any debris from this cavity.

Remove and clean the pump impeller.

Clean the foam block in a bucket of pond water to keep any beneficial bacteria alive.

Rinse the plastic casing in fresh water to remove debris.

SEASONAL POND MAINTENANCE

As the seasons change, your pond needs differing levels and types of maintenance. These include removing leaves in autumn, cleaning filter media regularly in summer and adding a pond heater in winter. This is a practical guide.

Spring is the time when a pond comes to life after a period of inactivity throughout the winter. Plants begin to grow, fish become more active, insect life increases and toads and frogs may begin to appear. During this period, when the pond water is warming up, the climate is subject to fluctuations that can prove stressful to fish, whose immune systems are not yet fully working. In temperate climates, nights may remain cold and heavy frosts can lead to a rapid drop in water temperature.

Spring colour

Caltha palustris, the marsh marigold, brightens the pond edge in spring and heralds the beginning of a new season of pond activity.

Fish will begin to look for food as the temperature increases, but it is important not to overfeed them, as biological filters will be slow to start functioning efficiently after the winter period. When feeding after the winter, make sure that the fish eat all the food you offer them. Do not let the food sink and decay, as biological activity will not yet be sufficient to cope with the increased bioloading. There are several foods available that are designed specifically for low-temperature feeding, and they provide an easily digested diet while activity in the pond is still slow in early spring.

Once water temperatures stabilize at over 12°C (54°F) throughout the day, you can introduce a standard staple food. At this temperature, the fishes' metabolism will increase and will work more efficiently at digesting food.

If filters and pumps were not in operation in winter, now is the time to reinstall them and turn them on. Use a bacteria culture to kick-start filters as the temperature rises. If you have been using a pond heater, remove and store it until the following year.

Testing the water

It is important to test for ammonia and nitrite throughout early spring. The increase in fish metabolism, coupled with regular feeding, can lead to a rapid increase in these toxins. In addition, plant matter that has died during the winter will decompose as the water temperature increases. Biological filters take time to establish again after the winter and are thus less effective at dealing with increased biological activity.

If you detect ammonia or nitrite levels, carry out a partial water change of between 10 and 20%. Remember to add a dechlorinator if using tapwater. A biological filter aid can be added if nitrite levels persist.

Above: Testing pond water with a paper strip test is very easy. Dip a strip into the water, wait one minute and then compare the colours with the chart provided. Each strip tests for nitrate, nitrite, total hardness, carbonate hardness and pH.

Alternatively, zeolite can be added to remove ammonia, although this should be seen as a short-term measure. If nitrite levels rise, reduce feeding until the filter is mature enough to cope with the additional biological requirement.

Checking pond equipment

Ideally, pumps should be allowed to run at a reduced flow rate throughout the winter. With the arrival of spring, increase the flow of water through the filter. Now is also the time to check that pumps and other electrical items are working correctly. Clean equipment, such as pumps and ultraviolet

The pond in spring

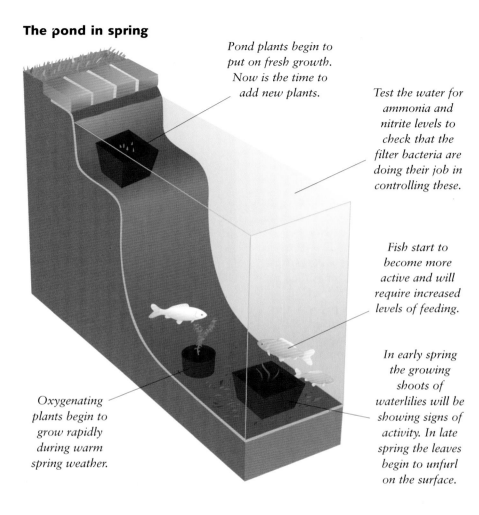

Pond plants begin to put on fresh growth. Now is the time to add new plants.

Test the water for ammonia and nitrite levels to check that the filter bacteria are doing their job in controlling these.

Fish start to become more active and will require increased levels of feeding.

In early spring the growing shoots of waterlilies will be showing signs of activity. In late spring the leaves begin to unfurl on the surface.

Oxygenating plants begin to grow rapidly during warm spring weather.

sterilizers, to ensure that winter debris does not affect their performance. Ultraviolet sterilizers should be turned on again to prevent early algae growth caused by the sudden increase in sunlight hours and intensity. Take this opportunity to replace UV lamps, which are generally only effective for six to twelve months. If pumps have been turned off through the winter, make sure that impellers, seals and other perishable parts are in good working order. Replace them if they require it.

Spring is also a good time to clean out a pond (see pages 110-121), but make sure that the fish are not spawning before you start work. Providing there is no spawning activity, fish and plants will cope with a clean out far better in early spring than at other times of the year.

Above: Long hours of sunlight, high levels of nitrates and a lack of plant life in the pond can lead to a build-up of blanketweed and other algae in spring.

Take plant cuttings and replant any specimens that require larger pots. Ponds should not require cleaning out every year. Rather, it is better to monitor the levels of silt and mud at the bottom of the pond on a regular basis and when these have built up so that the pond no longer has its usual spring 'sparkle' it may be time for an overhaul.

Right: Spring is the season when both plant and animal life begin to flourish. Waterlily leaves begin to unfurl at the surface and tadpoles make their appearance. If the pond has been netted for the winter, remove the net in good time for frogs to gain access to the pond for spawning.

Checking the fish

Providing the pond was correctly prepared for the winter period, fish and plants should enter springtime with minimal problems. It is always worth adding a general broad-spectrum treatment for fish in early spring to protect them while their immune systems are low.

Take time to inspect fish for abrasions and other wounds incurred during the winter. Now is the time to treat these before bacteria and other pathogens take advantage of the fishes' lowered immunity to disease.

Summer maintenance

Summer is a frantic time in the pond. Plants are in flower, fish and amphibians are spawning, insect life is at its peak and fish are feeding at their maximum level. Increased daylight and strong sunshine can lead to algae problems, and biological filters may buckle under the strain of a pond fully stocked with well-fed fish. Evaporation rates are highest in summer. Remember to

The pond in summer

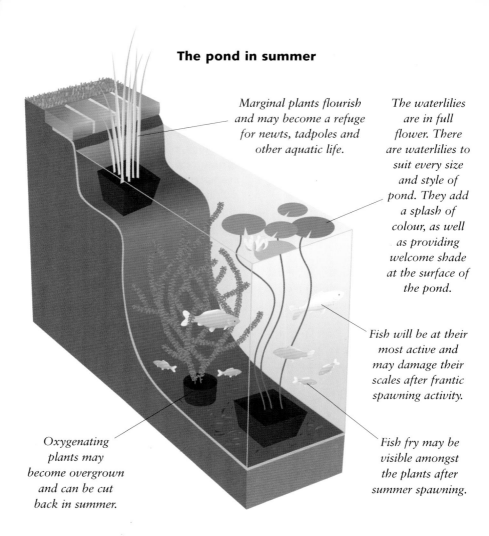

Marginal plants flourish and may become a refuge for newts, tadpoles and other aquatic life.

The waterlilies are in full flower. There are waterlilies to suit every size and style of pond. They add a splash of colour, as well as providing welcome shade at the surface of the pond.

Fish will be at their most active and may damage their scales after frantic spawning activity.

Oxygenating plants may become overgrown and can be cut back in summer.

Fish fry may be visible amongst the plants after summer spawning.

treat the replacement water before adding it to the pond. Use a proprietary dechlorinating chemical or a carbon filter to render tapwater safe for pond inhabitants.

Feeding in summer

Fish can be moved from low-protein, low-temperature foods onto a staple diet. Feed them two or three times each day, but only feed enough so that all the food is eaten within five minutes. Make sure that food is not sinking uneaten to the bottom of the pond. If bottom-feeding fish are present, feed them sparingly with sinking pellets, rather than relying on excess stick or flake food reaching them. Consider feeding live foods, such as daphnia and bloodworm.

Monitoring the water quality

Continue to test for ammonia and nitrite on a regular basis. This will give an indication of how the biological filtration is coping with the increased metabolism of the fish. If nitrite levels are consistently present, you may have to reduce the number of fish in the pond or increase the filtration.

Maintaining pond equipment

Clean prefilters, sponges and brushes on a regular basis to prevent filters from clogging and to ensure a steady and constant flow of water through them. Do not wash biological filter media in tapwater. Clean them in a bucket of pond water, which will ensure that beneficial bacteria are not destroyed in the cleaning process.

Above: Using a fountain or waterfall during warm summer months not only provides the welcome sound and sight of moving water, but also helps to keep the oxygen levels high.

Right: A healthy pond at the height of summer is a feature to grace any garden. Clear water, active fish and lush plant growth are the aim of every pond owner.

Oxygen levels

As temperatures rise, the oxygen content in the pond may drop to dangerously low levels. If this happens, the fish will start to gasp at the surface of the pond and can die as a result of lack of oxygen. Leave fountains and waterfalls running constantly during periods of hot weather to ensure that plenty of dissolved oxygen is made available. This is particularly important at night, when oxygen is no longer being produced by plants during photosynthesis. If a fountain or

waterfall is not present, use a pond airpump to add air to the pond and thus increase dissolved oxygen levels. During hot weather, installing a pond airpump near the pond adds much-needed oxygen to the water. Although a certain amount of oxygen passes into the water directly from the bubbles, an even greater oxygenating effect occurs at the surface of the churning water. Protect these pumps from the weather and use a ball-type airstone to provide a steady stream of air bubbles.

Plant care

Monitor pond plants closely and remove dead or dying leaves and flowers to prevent the risk of disease. Just like garden plants, pond plants are subject to pests, which are at their height during the summer months (see pages 66-69). Do not use general garden pesticides on pond plants and be careful not to allow pesticide sprays used in other areas of the garden to fall onto the pond surface, where they can poison the water and the fish.

Use a good-quality plant growth food during the summer. Suitable fertilizers are available in liquid, tablet or sachet form. Follow the suppliers' directions for use.

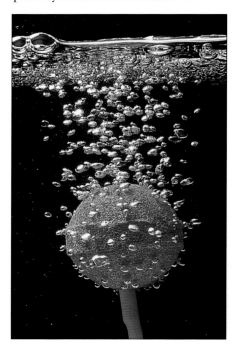

Above: During hot periods it is advisable to set up an airpump near the pond. The surface agitation caused by the rising bubbles increases gaseous exchange and raises oxygen levels in the water.

The pond in autumn

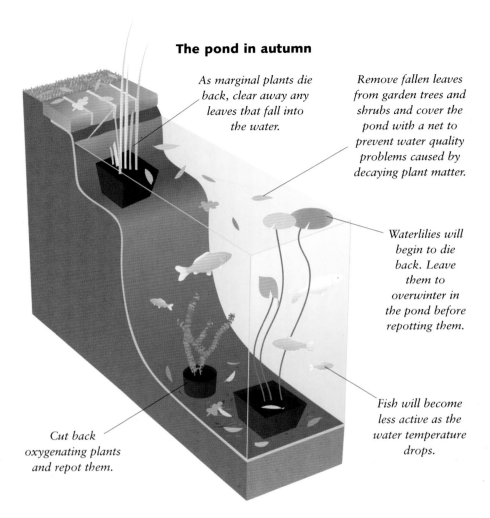

As marginal plants die back, clear away any leaves that fall into the water.

Remove fallen leaves from garden trees and shrubs and cover the pond with a net to prevent water quality problems caused by decaying plant matter.

Waterlilies will begin to die back. Leave them to overwinter in the pond before repotting them.

Fish will become less active as the water temperature drops.

Cut back oxygenating plants and repot them.

Autumn maintenance

Feed fish for as long as they are active through autumn months. Make sure that food is being eaten and not left to decompose. Start to introduce a good-quality, low-temperature food into the feeding regime. Check the fish for signs of disease and skin damage and treat them as appropriate. Treatments become ineffective as temperatures drop, so ensure that your fish are in peak condition before winter sets in. It is not a good idea to buy new plants in autumn, as they will soon die back as the temperature drops, so avoid the temptation of buying stocks that are often offered in late autumn sales.

Plant care

Autumn is the time to cut back plants as they die down. However, do not do this too early, as many aquatic plants will continue to grow and flower until the first autumn frosts in temperate areas. Some plants can even keep growing while there is ice on the pond.

Remove dead leaves and stems to prevent them decaying in the pond. Cut back oxygenating plants and repot them to avoid them dying during the winter. Keep the pond as free as possible from fallen leaves; add a pond cover net if necessary.

Pond equipment

Overhaul filters and pumps in autumn. Clean mechanical filter media but be careful not to wash beneficial bacteria away from biological media. Perform regular maintenance on UV sterilizers as recommended by the manufacturer and make sure that pumps are free from silt and other organic build-up.

Pump flows can be reduced in late autumn. To maintain the water temperature, it is important to turn

Above: Protect the pond from wind-blown leaves by covering the surface with a net. These are lightweight and easy to install by stretching them across the pond and securing the edges with plastic pegs.

Left: A pond margin in late autumn, showing the usual mix of fallen leaves and broken stems. Keep the water as clear as possible to reduce pollution problems. Even in the midst of this brown scene there are green splashes to hint at the new growth that will start up in spring.

The pond in winter

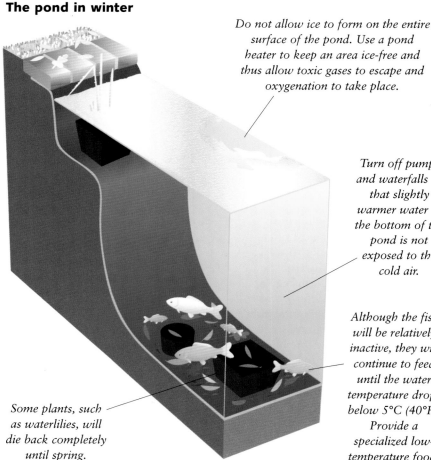

Do not allow ice to form on the entire surface of the pond. Use a pond heater to keep an area ice-free and thus allow toxic gases to escape and oxygenation to take place.

Turn off pumps and waterfalls so that slightly warmer water at the bottom of the pond is not exposed to the cold air.

Although the fish will be relatively inactive, they will continue to feed until the water temperature drops below 5°C (40°F). Provide a specialized low-temperature food.

Some plants, such as waterlilies, will die back completely until spring.

off fountains and waterfalls. If they are left running, they take warm water from the pond and expose it to cold air. This can have a dramatic effect on the water temperature and should therefore be avoided.

Winter maintenance

As water and air temperatures reduce, life in the pond begins to slow down. Fish will conserve energy through the winter months by lying at the bottom of the pond, where the water is warmer than at the top. Use a pond thermometer to determine the temperature of the water, as it is often above 5°C (40°F), even when the air temperature is far lower. At temperatures below 5°C (40°F), fish enter a state of semi-hibernation and should not be fed. However, at above this temperature, you should continue to feed them with a low-temperature food.

Fish care

Most low-temperature foods contain wheatgerm, which is the most easily digested vegetable protein. However, the best-quality foods are based on fish protein, as this is far more easily digested by pond fish and enables them to continue growing throughout the winter months. Carefully read the ingredients on low-temperature pond foods before buying them to ensure that your fish receive the best possible diet throughout the winter.

During the winter, when plant cover is at its lowest, fish are susceptible to attack from herons and other natural predators. Use a pond cover net or decoy heron to discourage predators. Cover nets will also help to keep fallen leaves off the pond surface.

Pond equipment

Fountains and waterfalls should be turned off by now. Raise up pumps from the base of the pond to leave a layer of warmer, undisturbed water at lower levels. If pumps have been turned off, remove them from the pond and check that all pipework is free of water to protect it from ice damage as the temperature drops in temperate climates. This also applies to filters if they are not being run all year round. If you are leaving pumps in the pond, run them occasionally to prevent calcium build-up and

Left: Use a pond heater or ice vent to ensure that at least part of the pond remains ice-free. Allowing the pond to ice over completely can cause a build-up of dangerous gases.

clogging from silt. If external pumps are not to be used continuously, you must drain and store them according to the manufacturer's instructions.

Protecting the pond from icing up

Do not allow ice to cover the entire pond surface as this can lead to problems. As ice covers the pond, it prevents the escape of harmful gases and stops oxygenation taking place. This can kill fish and it is therefore important to keep an area of the pond ice-free. Do not break the ice, as shock waves can harm or even kill fish. Instead, use either a pond heater or a polystyrene foam ice vent to keep an area ice-free. If the pond does become covered in ice, use a small pan of hot water to make a small opening into which you can place a heater. The purpose of such heaters is not to warm the pond water, but to keep a small patch of the pond free of ice.

It is also important to prevent ice from damaging ponds. As ice forms, it exerts great pressure on the sides of a pond. Although preformed ponds are usually of sufficient quality to absorb the pressure, concrete ponds certainly are not. Using a polystyrene foam ice vent will relieve some of the pressure. Do not use plastic balls or containers, as they do not allow gaseous exchanges to take

Using an ice vent

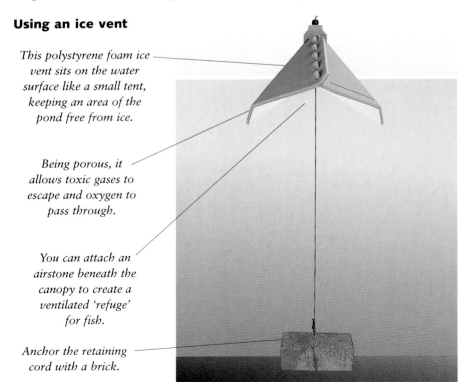

This polystyrene foam ice vent sits on the water surface like a small tent, keeping an area of the pond free from ice.

Being porous, it allows toxic gases to escape and oxygen to pass through.

You can attach an airstone beneath the canopy to create a ventilated 'refuge' for fish.

Anchor the retaining cord with a brick.

Left: A pond heater will warm up a small patch of water, sufficient to keep an area ice-free and allow gaseous exchange to take place. In a small pond, as shown here, a heater may keep the entire surface ice-free unless conditions become very cold. The fish tend to congregate in the zone of warmth the heater creates.

place. Most ice vents are made from polystyrene foam, which is porous enough to allow the required amount of gaseous exchange, and will expand and contract, thus providing protection from ice damage. Nevertheless, polystyrene foam vents will still freeze in very low temperatures and should therefore be used in conjunction with a heater or airpump.

Finally, continue to remove floating debris from the pond throughout the winter to prevent water quality problems occurring as the temperature rises in spring.

Above: When selecting a pond heater, choose one with a polystyrene foam collar to ensure that it floats. This makes regular maintenance checks more convenient.

INDEX

Page numbers in **bold** indicate major entries, *italics* refer to captions and annotations; plain type indicates other text entries.

A

nitrate *17*, *123*
nitrite 16, *17*, 123, *123*
paper test strip *123*
pH 18, *18*, *123*
Tinca tinca 78
Toads 122
Treatments
  antibacterial 100,
    101, 102, 103
  anti-parasite pond
    97, *97*, 99, 101, 102
  anti-ulcer remedy 104
  broad spectrum 125
    antibacterial 105
  proprietary anti-
    fungus 108
  tonic salt 104, 105
  whitespot 98
Turions 60
  butomus *59*
  winter 60

U

Ultraviolet (UV)
  sterilizer 21, 40, *42*,
    43, 44, 45, 112,
    123, 124, 131
  choosing 43
  housing *42*
  linear UV lamp *42*
  'PL' UV lamp 43, *43*
  quartz sleeve *42*, 43, 44

submersible unit 44
underwater *43*

V

Veronica beccabunga 62

W

Waterfalls 20, 21, 23, 24,
  26, *32*, *34*, *35*, *36*, 127,
  128, 132, 133
Water fleas 26
Waterlilies 38, 51, 52,
  55, 56, 65, 80, *124*,
  125, *126*, *128*, *130*
  aphids 65, 66
  bare-rooted 51
  beetles 66, 67
  crown rot 69
  division 56, *56*, 57
  hardy 57, *57*
  pots 116
  preparing and planting
    51
  propagating *58*
  pygmy 56
  repotting 119
  small-growing 56
Water mint 62
Water plantain 56
Water quality 11, **12-21**,

41, 72, *96*, 97, 101,
  130
acid-alkaline balance 12
acidity 13, 18, 19
algicide treatment 19, *41*
alkalinity 13, 18, 19
  buffer 20
ammonia levels 16, *96*,
  124, 126
bacteria levels 12
carbonate hardness 123
chlorine 12, 13, 114,
  117
dechlorinator, 13, *117*,
  123, 126
evaporation 125
nitrite levels *96*, 126
oxygen levels 12, 20,
  20, 21, *96*, 127, 128
pH levels *96*, 123
poor 38
problems *45*, 47
tap water 12, 13
  conditioner *13*, *114*
  purifier 13
testing (*see* Test kits)
total hardness 123
understanding pond
  water 13
Water soldier 60

Z

Zeolite 23, 28, 30

CREDITS

The practical photographs featured in this book have been taken by Geoffrey Rogers and are © Interpet Publishing.

The publishers would like to thank the following photographers for providing images, credited here by page number and position: B(Bottom), T(Top), C(Centre), BL(Bottom Left), etc.

David Allison: 79(B)
James Allison/Aquapic: 21
MC & C Piednoir/Aqua Press - France: 22, 23, 38, 39, 74-75, 76-77(B), 79(T), 83, 86, 90(C), 94(Peter Cole), 95, 104, 106
Biofotos/Heather Angel: 51, 81(T), 98, 108(T), 127(B)
Eric Crichton: 49, 57
The Garden Picture Library: Intro page (John Glover)
John Glover: 129
Harpur Garden Library: 40-41(B, Ernie Taylor)
Andrew Lawson: 50
S & O Mathews: 122, 127(T)
Clive Nichols: Credits page (designer Richard Coward), 12 (Mr Fraser/J. Treyer-Evans)
Oxford Scientific Films: 93(Max Gibbs), 107(Max Gibbs)
PHOTOMAX: 70, 77(T), 84-5(B), 103(T), 105
Photos Horticultural: 66(B)
Geoffrey Rogers: 44, 55, 67, 68, 87, 110,111, 125, 128, 131, 133, 135
Fred Rosenzweig: 76(T)
Mike Sandford: 80-81(B), 85(TR), 96, 100, 103(B), 109(B)
Neil Sutherland © Geoffrey Rogers: Title page
David Twigg: 82

The artwork illustrations have been prepared by John Sutton and are © Interpet Publishing.

ACKNOWLEDGMENTS

Thanks are due to Blagdon Garden Products Ltd., Bridgewater, Somerset; John Buery, New Ash Green, Kent; The Dorset Water Lily Company, Halstock, Dorset; Heaver Tropics, Ash, Kent; Phoenix 2000, Pinxton, Nottinghamshire; Rotorflush Filters, Charmouth, Bridport, Dorset; Tetra UK, Eastleigh, Hampshire; June Crowe at Rose Cottage, Hartley, Kent.

The information and recommendations in this book are given without any guarantees on the part of the author and publisher, who disclaim any liability with the use of this material.